D1578776

Unit C1: Chemistry in Our World

Unit C2: Discovering Chemistry

Unit C3: Chemistry in Action

C1 The Earth's Sea and Atmosphere

C1 Topic 1: The Earth's Sea and Atmosphere

This topic looks at:

- how the early atmosphere and oceans were formed
- how levels of carbon dioxide reduced and levels of oxygen increased
- the impact of volcanic and human activities on the atmosphere

History of the Atmosphere

Since the formation of the Earth 4.6 billion years ago, the atmosphere has changed dramatically.

The timescale, however, is enormous: one billion years is one thousand million (1 000 000 000) years.

Timescale	Composition of the Atmosphere	Key Factors and Events that Shaped the Atmosphere
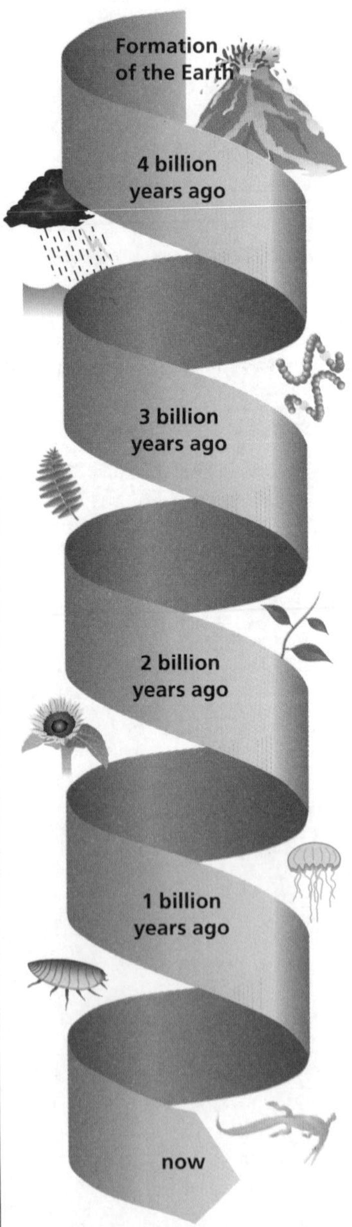 Formation of the Earth 4 billion years ago 3 billion years ago 2 billion years ago 1 billion years ago now	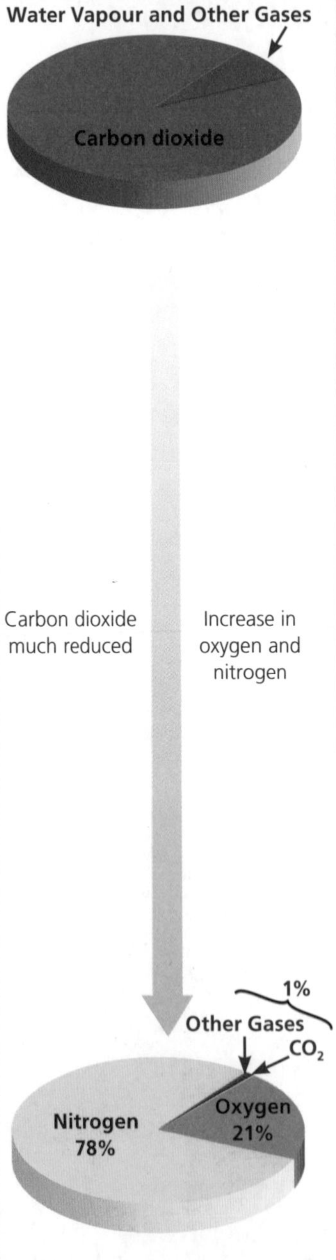 Water Vapour and Other Gases Carbon dioxide Carbon dioxide much reduced Increase in oxygen and nitrogen 1% Other Gases CO_2 Oxygen 21% Nitrogen 78%	Volcanic activity released carbon dioxide and small amounts of ammonia, methane and hydrogen. This volcanic activity also produced water vapour, which condensed as the Earth cooled, falling as rain and filling up the hollows in the crust to form oceans. Primitive green plants such as single-celled algae evolved and the amount of carbon dioxide was reduced as the plants took it in and gave out oxygen during photosynthesis. Carbon from carbon dioxide in the air was locked up in sedimentary rocks as carbonates and fossil fuels. The carbonates were formed when carbon dioxide dissolved in the oceans (see page 5). There is now more oxygen in the atmosphere. Some is converted to ozone, protecting living organisms from the Sun's ultraviolet radiation and allowing new species to evolve. There is a state of approximate balance because: • photosynthesis produces oxygen in the presence of sunlight and uses carbon dioxide • respiration and burning fuels use oxygen and produce carbon dioxide • carbon dioxide is absorbed by the seas and oceans.

Scientific Evidence About the History of the Atmosphere

Scientists have different sources of information about the development of the Earth's atmosphere, which means they are unable to be precise about the evolution of the Earth's atmosphere.

Scientists do have the following evidence, which has helped them to conclude that there was little or no oxygen in the early atmosphere and that the amount of oxygen within the atmosphere has increased over time:

- Oxygen is not released by volcanoes.
- Venus and Mars have similar atmospheres made up of carbon dioxide.
- Fossil evidence from early rock shows the existence of photosynthesising organisms.
- The composition of iron compounds is different within different types and ages of rock.

Carbon Dioxide in the Atmosphere

When the Earth was young, the atmosphere was mostly made up of carbon dioxide. Gradually this decreased due to photosynthesis of primitive plants converting the carbon dioxide and water to glucose and oxygen. The amount of carbon dioxide in the atmosphere also decreased as it dissolved in the newly formed oceans.

Some of the dissolved carbon dioxide reacted with other substances in the sea to form insoluble compounds such as **calcium carbonate** and soluble compounds such as **calcium hydrogencarbonate**. These compounds became concentrated in the shells of sea creatures. When the sea creatures died their remains eventually formed carbonate rocks, locking away the carbon.

The levels of carbon dioxide in the atmosphere are generally thought to have been in balance over the last 200 million years.

The constant recycling of carbon is called the **carbon cycle.**

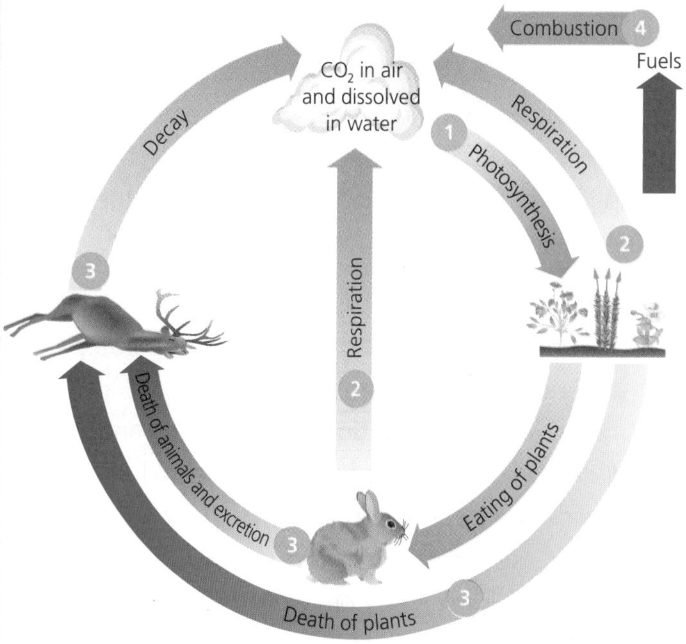
Combustion 4
Fuels
CO_2 in air and dissolved in water
Decay
Respiration
1 Photosynthesis
2
3
Respiration
2
Death of animals and excretion 3
Eating of plants
Death of plants
3

1. Carbon dioxide is removed from the atmosphere by green plants during photosynthesis.
2. Plants and animals respire, releasing carbon dioxide into the atmosphere.
3. Microorganisms feed on dead plants and animals, causing them to break down, decay and release carbon dioxide into the air. (The microorganisms respire as they feed.)
4. The burning of fossil fuels also releases carbon dioxide into the air.

Present day volcanic eruptions contribute to the levels of carbon dioxide within the atmosphere. The burning of fossil fuels also releases carbon dioxide into the atmosphere. If any branch of the carbon cycle is changed, then this will impact on the level of carbon dioxide in the atmosphere, resulting in either an increase or a reduction.

For example, deforestation takes away the trees that convert the carbon dioxide into oxygen, so there will be an increase in levels of carbon dioxide.

Animals exhale carbon dioxide when they respire, so an increase in farming livestock increases the amount of carbon dioxide produced through respiration.

C1 The Earth's Sea and Atmosphere

Global Warming

Global warming refers to the gradual increase in the Earth's temperature that has occurred. This can happen when greenhouse gases, such as carbon dioxide, water vapour, nitrous oxide and methane, trap energy from the Sun in the Earth's atmosphere, which increases the temperature. The level of greenhouse gases in the atmosphere has gradually increased as a result of:

- **burning fossil fuels** (oil, coal, gas), which produce carbon dioxide, sulfur dioxide (which contributes to **acid rain**) and carbon monoxide
- motorised **transportation** burning petrol and diesel, which produces carbon dioxide, sulfur dioxide and carbon monoxide
- an increase in **cattle farming and rice growing** – methane is released from wetlands (where rice grows) and from animals (particularly cattle)
- **deforestation** – when trees grow, they take in carbon dioxide. If more and more trees are cut down, less carbon dioxide is removed from the atmosphere.

Oxygen in the Atmosphere

The levels of oxygen in the atmosphere have changed over the history of the Earth, rising steadily from non-existence about 2 billion years ago, to about 21% today. Scientists believe it was the evolution of primitive plants such as algae that caused the atmosphere to change and the carbon dioxide to be used up during photosynthesis. This process produced the first molecules of oxygen.

Carbon dioxide + Water $\xrightarrow{\text{Light}}$ Glucose + Oxygen

HT $$6CO_2 + 6H_2O \xrightarrow{\text{Light}} C_6H_{12}O_6 + 6O_2$$

Measuring Oxygen in the Atmosphere

A simple practical demonstration can be carried out to show the level of oxygen present in the atmosphere.

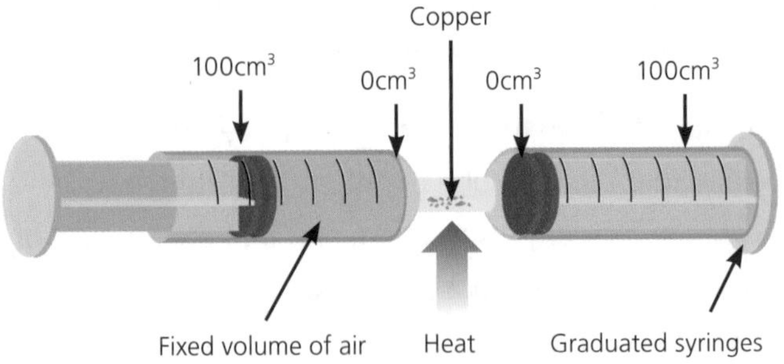
Copper
100cm³
0cm³
0cm³
100cm³
Fixed volume of air
Heat
Graduated syringes

If copper is heated while a fixed volume of air passes over it, then the volume of air will gradually reduce as the oxygen is used up in the chemical reaction. Eventually the volume of air will stop changing because all the oxygen will have been used up.

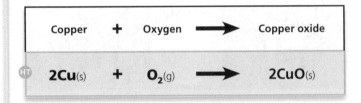

Copper + Oxygen $\longrightarrow$ Copper oxide

HT $$2Cu(s) + O_2(g) \longrightarrow 2CuO(s)$$

Volume of Air during the Course of the Experiment

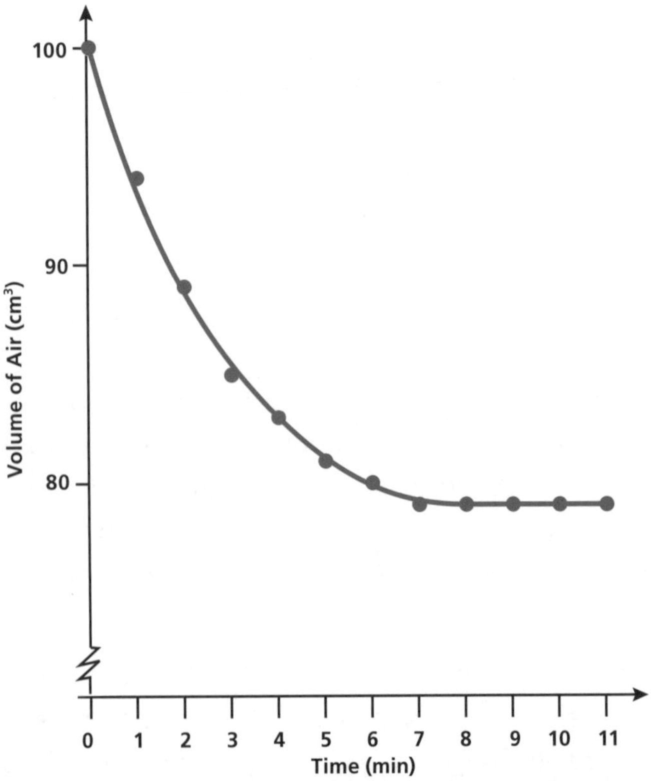
100
90
80
Volume of Air (cm³)
0
1
2
3
4
5
6
7
8
9
10
11
Time (min)

C1 Topic 2: Materials from the Earth

This topic looks at:
- how igneous, sedimentary and metamorphic rocks are formed
- everyday uses of limestone (calcium carbonate) and their environmental impacts
- what is involved in the thermal decomposition of carbonates

Different Types of Rock

Rocks are mixtures of **minerals**. A mineral is any solid element or compound that can be found in the Earth's crust.

There are three types of rock:
- igneous
- sedimentary
- metamorphic.

Igneous rock is produced when molten rock cools down to form crystals.

Molten rock, or magma, that erupts from a volcano is called lava. When lava cools, it cools very quickly, forming small crystals. This type of rock is called **extrusive igneous rock**. An example of this type of rock is basalt.

If the magma does not reach the surface it can still cool down but will do so very slowly, forming large crystals. This type of rock is called **intrusive igneous rock**. An example of this type of rock is granite.

Granite

Sedimentary rock, such as chalk and limestone, is formed as a result of a build-up of layers of sediment.

The pressure (compaction) of the layers forces out water but leaves behind the minerals that had been dissolved in it. These minerals act like a cement and hold the particles together. The layers that make up sedimentary rock are where **fossils** can be found.

However, **sedimentary rock** is porous because there are tiny gaps between the particles. These tiny gaps make the rock susceptible to erosion because they are points of weakness.

Metamorphic rock forms as a result of rock being buried due to the movement of the Earth's crust. The deeper the rock is buried, the greater the heat and pressure to which it is exposed. Over time this will cause the rock to change without ever melting. The minerals in the rock will line up to form bands or sheets of tiny grains or crystals. If this happens to limestone or chalk then eventually marble is formed.

Marble

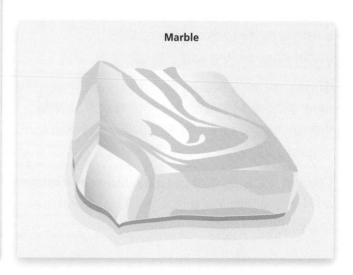

Limestone

Calcium carbonate, $CaCO_3$, can be found in rocks that exist in the Earth's crust as chalk, limestone and marble. Limestone is a very important building material and has industrial uses, for example, it is used in the extraction of iron from its ore (see page 14).

Quarrying Limestone

To extract limestone from the ground it must be **quarried**. This means digging a large hole and taking the rock away.

There are many factors that have to be considered before the limestone can be quarried. These factors include:

- the effect on the landscape and native animal habitats
- the effect on local businesses
- costs involved in quarrying the rock and then processing it
- whether there is an available local workforce
- the effect of noise pollution
- if there will be an increase in air dust and dirt levels
- if there will be additional traffic and new roads required
- how the quarry site can be used afterwards.

Uses of Limestone

As a building material, limestone is used in numerous ways.

It can be used straight from the quarry as block to construct the walls and floors of buildings. Many buildings built near limestone quarries are manufactured in this way.

Glass is made by heating limestone, $CaCO_3$, with sand, SiO_2, and sodium carbonate, Na_2CO_3. When they are heated they all melt together and then cool to form a transparent solid.

Cement is the result of heating powdered limestone and clay together in a rotary kiln. It is the heat that causes the two materials to react together.

Concrete is the result of mixing cement, sand, crushed rock or gravel and water together. As the mixture sets it forms a hard, stone-like solid. Steel supports can be used to make the concrete stronger. This is called reinforced concrete.

Acidic sulfur dioxide gas emitted by coal-fired power stations can be neutralised by using limestone or lime before the gas leaves the chimneys.

This reaction forms calcium sulfate, $CaSO_4$, which can be used in the building industry as **plaster**.

Farmers can **neutralise soil** that is too acidic to grow crops by spreading powdered limestone or lime onto the soil. Lime is another name for calcium oxide, CaO. Calcium oxide can be obtained from the thermal decomposition of calcium carbonate.

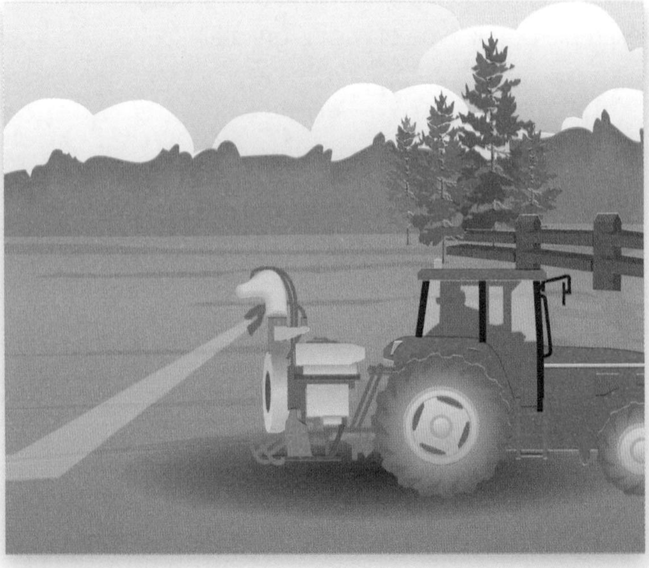

Thermal Decomposition of Carbonates

When calcium carbonate is heated strongly enough, it will break down and form different molecules. This process is called **thermal decomposition**. Thermal decomposition is very important for calcium carbonate as it produces calcium oxide, a very reactive material that has many uses, including in the neutralisation of acidic soils, as indicated on the previous page.

The equation below describes the thermal decomposition of calcium carbonate.

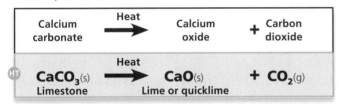

Calcium carbonate	Heat →	Calcium oxide	+	Carbon dioxide
HT $CaCO_3(s)$ Limestone	Heat →	$CaO(s)$ Lime or quicklime	+	$CO_2(g)$

The following equations show how limewater (used when testing for carbon dioxide gas) is produced from calcium oxide.

Calcium oxide	+	Water	→	Calcium hydroxide
HT $CaO(s)$ Quicklime	+	$H_2O(l)$ Small amount of water added	→	$Ca(OH)_2(s)$ Slaked lime

Calcium hydroxide	+	Water	→	Calcium hydroxide solution
HT $Ca(OH)_2(s)$ Slaked lime	+	$H_2O(l)$ Lots more water added	→	$Ca(OH)_2(aq)$ Limewater

Some carbonates will decompose more easily than others and this can be demonstrated in a simple investigation in the laboratory.

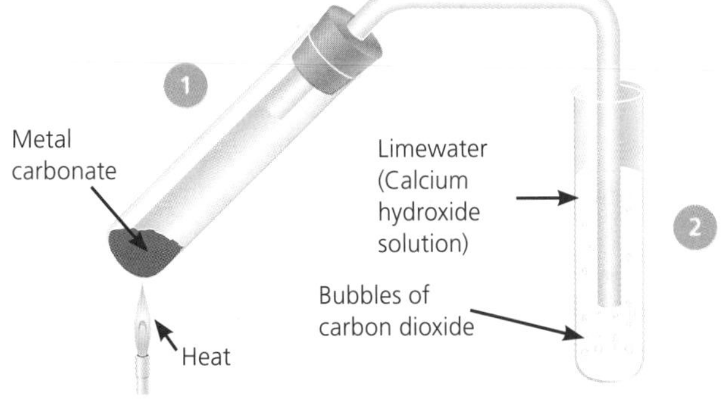
1
Metal carbonate
Heat
Limewater (Calcium hydroxide solution)
Bubbles of carbon dioxide
2

The more bubbles that appear in the limewater, the easier the carbonate is to decompose.

Bubbling the carbon dioxide through limewater means it will not all escape. It will also result in the limewater turning milky white or cloudy as calcium carbonate is precipitated.

For example, the equations below illustrate what happens during the decomposition of zinc carbonate:

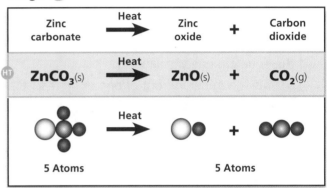

Stage 1

Zinc carbonate	Heat →	Zinc oxide	+	Carbon dioxide
HT $ZnCO_3(s)$	Heat →	$ZnO(s)$	+	$CO_2(g)$

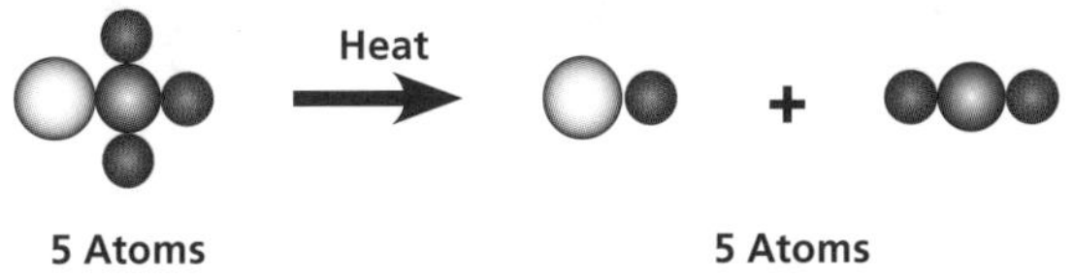
Heat
5 Atoms
5 Atoms

Stage 2

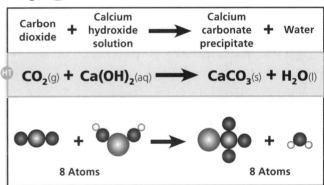

Carbon dioxide	+	Calcium hydroxide solution	→	Calcium carbonate precipitate	+	Water
HT $CO_2(g)$	+	$Ca(OH)_2(aq)$	→	$CaCO_3(s)$	+	$H_2O(l)$

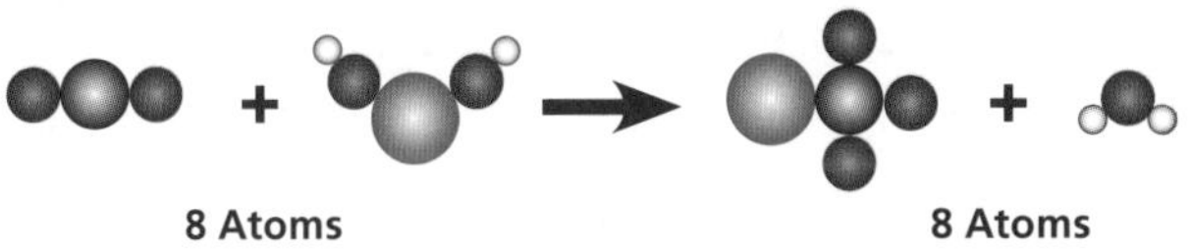
8 Atoms
8 Atoms

From the equation it can be seen that no atoms have been lost or gained, they have just rearranged themselves. **Conservation of mass** means that total masses at the start and finish will also have stayed the same.

Remember that **atoms** are the basic particles from which all matter is made. All chemical **elements** are made up of atoms.

Balancing Equations

All chemical reactions follow the same simple rule: the total mass of the reactants is equal to the total mass of the products.

This means there must be the same number of atoms on both sides of the equation.

Writing Balanced Equations

1. Write down the word equation for the reaction.
2. Write down the correct formula for each of the reactants and the products.
3. Check that there are the same numbers of each atom on both sides of the equation.

If the equation is already balanced, leave it.

If the equation needs balancing:

4. Write a number in front of one or more of the formulae. This increases the number of all of the atoms in the formula.
5. Don't forget the state symbols: (s) = solid, (l) = liquid, (g) = gas and (aq) = dissolved in water (aqueous solution).

Example

Balance the reaction between sodium and water.

1. The word equation for the reaction is:

Sodium + Water ⟶ Sodium hydroxide + Hydrogen

2. The correct formulae for each of the products and the reactants are:

$$Na + H_2O \longrightarrow NaOH + H_2$$

3. Check there are the same number of atoms on each side of the equation.

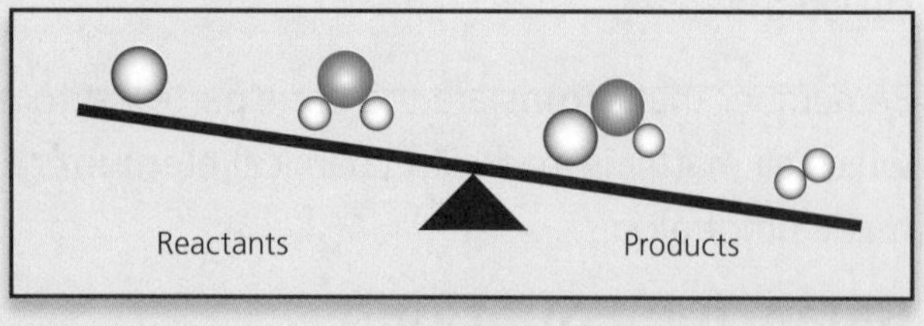
Reactants
Products

There are more hydrogen atoms on the products side than on the reactants side. The equation needs balancing.

4. Balance the hydrogen by doubling the amount of water and sodium hydroxide.

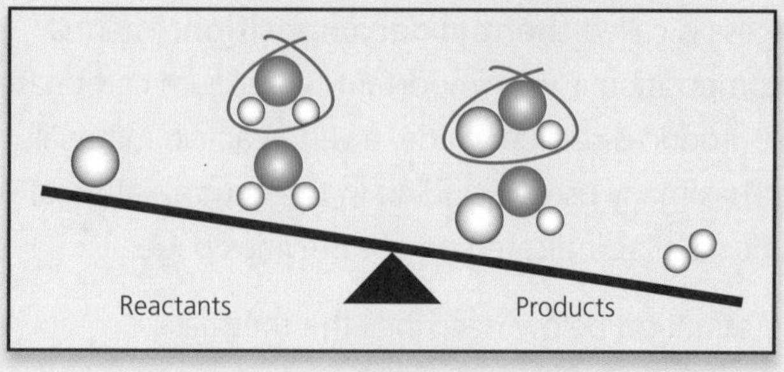
Reactants
Products

The amount of oxygen and hydrogen on both sides is equal. However, the amount of sodium is now unequal.

Double the sodium on the reactants side to match the amount of sodium on the products side.

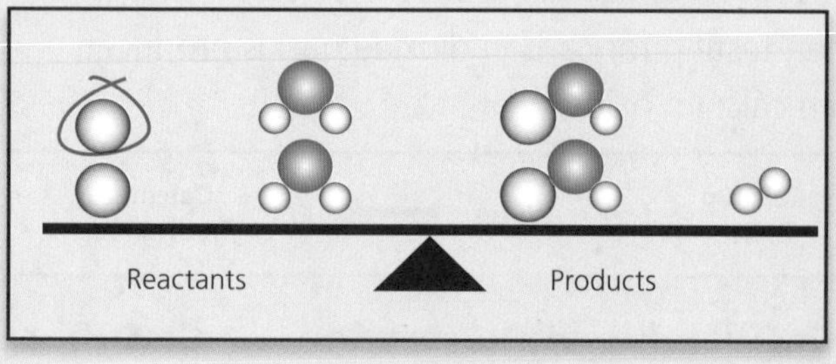
Reactants
Products

This gives you the balanced equation.

$$2Na + 2H_2O \longrightarrow 2NaOH + H_2$$

5. Don't forget to include the state symbols.

$$2Na_{(s)} + 2H_2O_{(l)} \longrightarrow 2NaOH_{(aq)} + H_{2(g)}$$

C1 Topic 3: Acids

This topic looks at:

- the importance of neutralisation reactions and the electrolysis of hydrochloric acid
- the neutralisation reactions of other acids
- the uses of chlorine and methods of producing it
- what happens during the electrolysis of water

Acids

An **acid** is a substance that produces **hydrogen ions** when in an aqueous solution, for example, hydrochloric acid (HCl), sulfuric acid (H_2SO_4) and nitric acid (HNO_3). Hydrogen ions can be identified by using an **indicator**, such as a universal indicator, which indicates a range of pH values.

pH Scale

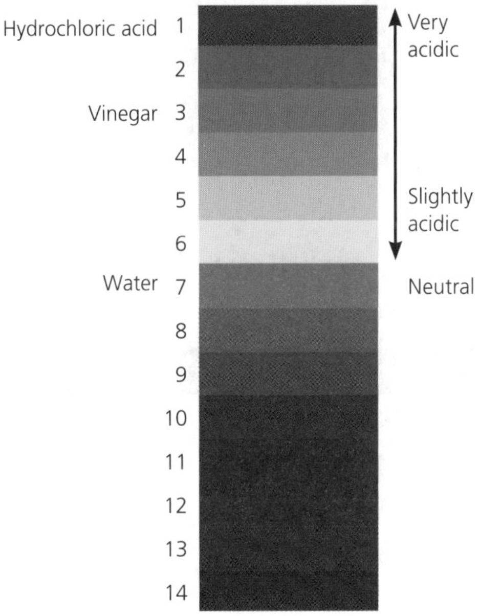

Hydrochloric acid 1
2
Vinegar 3
4
5
6
Water 7
8
9
10
11
12
13
14
Very acidic
Slightly acidic
Neutral

Making Salts

A neutral **salt** can be formed when an acid is reacted with a base. (Some bases are soluble in water and dissolve to produce alkaline solutions.) This type of reaction is called a **neutralisation** reaction. Most of the salts formed in these reactions are soluble and they can only be obtained when the water is evaporated off.

Using an Alkaline Hydroxide Base

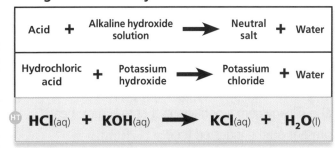

Acid	+	Alkaline hydroxide solution	→	Neutral salt	+	Water
Hydrochloric acid	+	Potassium hydroxide	→	Potassium chloride	+	Water

HT $$HCl_{(aq)} + KOH_{(aq)} \rightarrow KCl_{(aq)} + H_2O_{(l)}$$

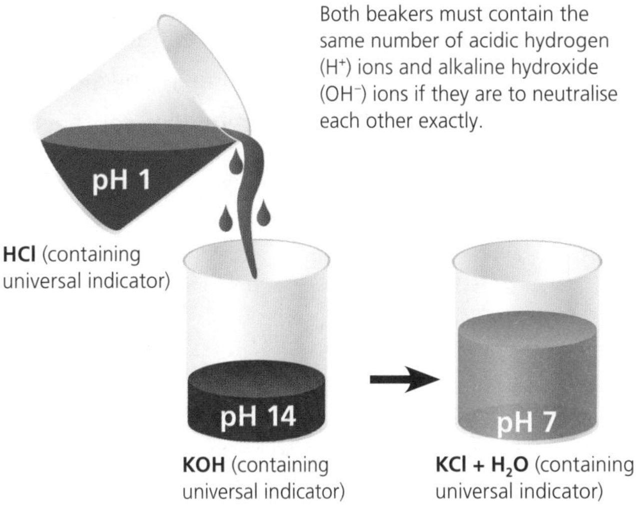

Both beakers must contain the same number of acidic hydrogen (H^+) ions and alkaline hydroxide (OH^-) ions if they are to neutralise each other exactly.
pH 1
HCl (containing universal indicator)
pH 14
KOH (containing universal indicator)
pH 7
KCl + H_2O (containing universal indicator)

Using a Metal Oxide Base

Acid	+	Metal oxide	→	Neutral salt	+	Water
Sulfuric acid	+	Copper oxide	→	Copper sulfate	+	Water

HT $$H_2SO_{4(aq)} + CuO_{(s)} \rightarrow CuSO_{4(aq)} + H_2O_{(l)}$$

Using a Metal Carbonate Base

Acid	+	Metal carbonate	→	Neutral salt	+	Water	+	Carbon dioxide
Nitric acid	+	Calcium carbonate	→	Calcium nitrate	+	Water	+	Carbon dioxide

HT $$2HNO_{3(aq)} + CaCO_{3(s)} \rightarrow Ca(NO_3)_{2(aq)} + H_2O_{(l)} + CO_{2(g)}$$

The particular salt produced depends on the acid and the metal in the base used. They all react in the same way but produce different salts. Sulfuric acid produces sulfate salts, hydrochloric acid produces chloride salts and nitric acid produces nitrate salts.

Everyday Neutralisation Reactions

Hydrochloric acid is used by the human body to digest food. It is an acid that is produced in the stomach to help break down food into smaller molecules. It will also kill any germs or bacteria that may have been present on the ingested food.

Sometimes too much acid is produced by the stomach and this can cause pain. Excess acid in the stomach can be neutralised by taking indigestion tablets. Each of these tablets or remedies contains a base that will neutralise the acid.

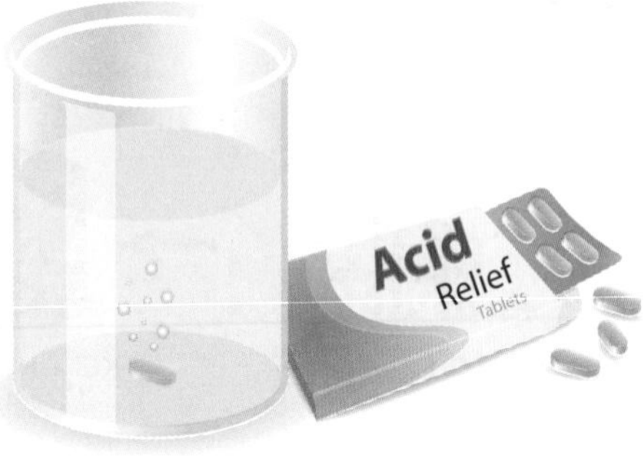
Acid
Relief
Tablets

Remember:

Acid	**+**	**Alkali**	**→**	**Salt**	**+**	**Water**

This can be tested in the laboratory by neutralising some hydrochloric acid with a range of different indigestion remedies.

Electrolysis

When electrical energy from a direct current (d.c.) supply is passed through a compound that is molten or in solution, the compound is **decomposed**. The ions move to the electrode of opposite charge.

The ions that are positively charged move towards the negative electrode.

The ions that are negatively charged move towards the positive electrode.

When the ions get to the electrodes, they are **discharged** (lose their charge) and atoms of elements are formed.

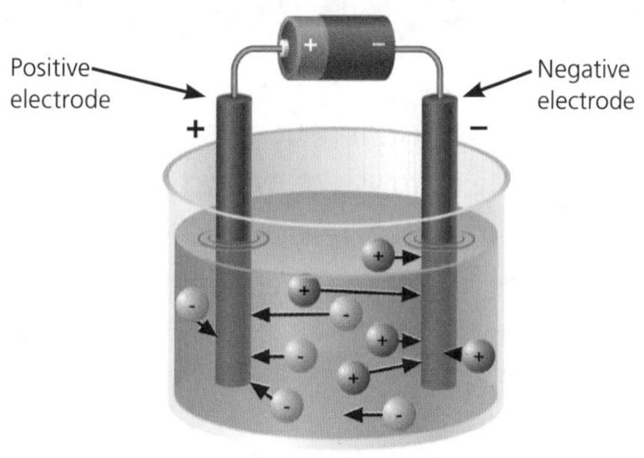
Positive electrode
Negative electrode
+
–

Electrolysis of Hydrochloric Acid

Electrolysis can be demonstrated in the laboratory by passing a current through dilute hydrochloric acid.

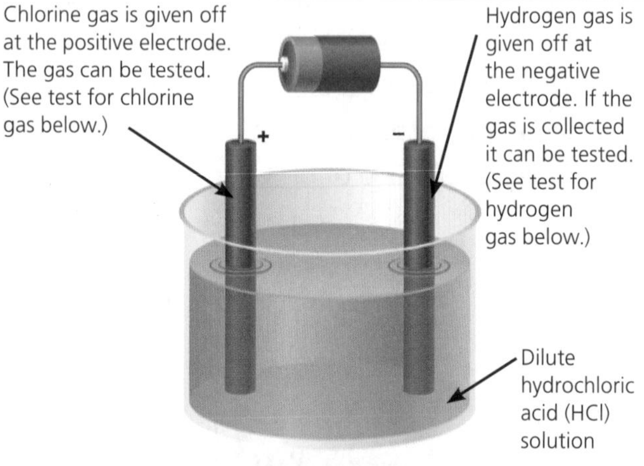
Chlorine gas is given off at the positive electrode. The gas can be tested. (See test for chlorine gas below.)
Hydrogen gas is given off at the negative electrode. If the gas is collected it can be tested. (See test for hydrogen gas below.)
+
–
Dilute hydrochloric acid (HCl) solution

The following table shows how to test for hydrogen gas and chlorine gas.

Gas	Properties	Test for Gas
Hydrogen, H_2	A colourless gas. It combines violently with oxygen when ignited.	When mixed with air, burns with a squeaky pop. 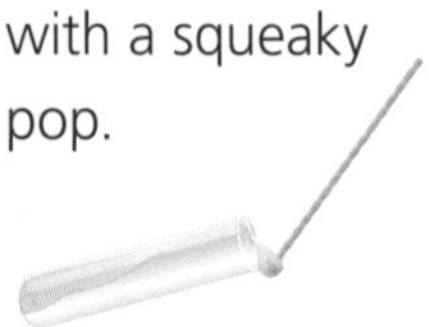
Chlorine, Cl_2	A green poisonous gas that bleaches dyes.	Turns damp indicator paper white. 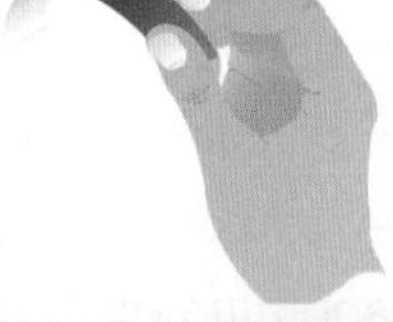

Electrolysis of Water

The electrolysis of water will result in its decomposition to hydrogen gas at the negative electrode and oxygen gas at the positive electrode.

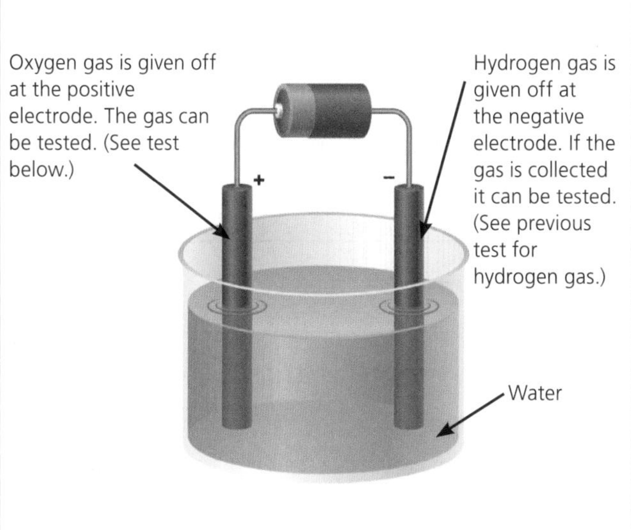
Oxygen gas is given off at the positive electrode. The gas can be tested. (See test below.)
+
–
Hydrogen gas is given off at the negative electrode. If the gas is collected it can be tested. (See previous test for hydrogen gas.)
Water

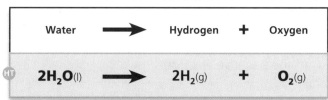

Water → **Hydrogen** + **Oxygen**

HT $2H_2O_{(l)} \rightarrow 2H_{2(g)} + O_{2(g)}$

The presence of these gases can be tested by using the test described on page 12 for hydrogen gas and the test for oxygen shown below.

Gas	Properties	Test for Gas
Oxygen, O_2	A colourless gas that helps fuels burn more readily than in air.	Relights a glowing splint. 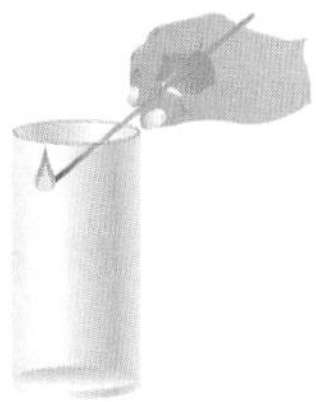

The Production of Chlorine by Electrolysis

Chlorine gas can be obtained during the industrial electrolysis of brine. Brine is the name that is given to water containing large amounts of salt (sodium chloride), such as sea water. Extra safety precautions must be taken by a company when chlorine gas is obtained on such a large industrial scale because it is a toxic gas.

When a concentrated solution of sodium chloride is electrolysed, chlorine gas is produced at the positive electrode, hydrogen gas is produced at the negative electrode and the solution that remains is the alkali sodium hydroxide.

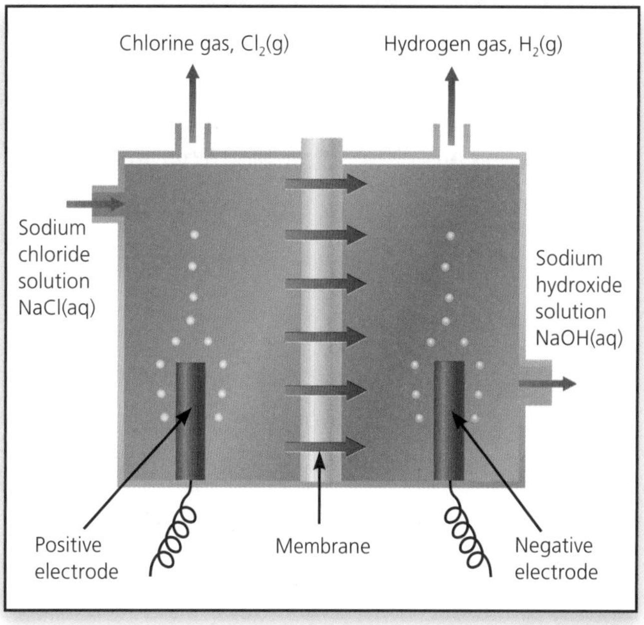
Chlorine gas, $Cl_2(g)$
Hydrogen gas, $H_2(g)$
Sodium chloride solution NaCl(aq)
Sodium hydroxide solution NaOH(aq)
Positive electrode
Membrane
Negative electrode

Sodium chloride + **Water** → **Hydrogen** + **Chlorine** + **Sodium hydroxide**

HT $2NaCl_{(aq)} + 2H_2O_{(l)} \rightarrow H_{2(g)} + Cl_{2(g)} + 2NaOH_{(aq)}$

Uses of Chlorine

Chlorine is a toxic gas that will kill harmful bacteria when added to water. It can be used to make bleaches and other disinfectants and can also be used to make the polymer poly(chloroethene) (PVC) (see page 23).

ULTRA
CLEAN
Drain
Cleaner
BLEACH
KILLS ALL GERMS - FAST!
ULTRA
WASH
NON-BIOLOGICAL

C1 Topic 4: Obtaining and Using Metals

This topic looks at:
- how metals are obtained from their ores
- what reduction and oxidation reactions are
- how metals are disposed of
- how properties determine the use of a metal
- how the reactivity series can predict extraction methods and reactions of metals
- how the properties of an alloy differ from the original metal

Metal Ores

Ores are naturally occurring rocks found in the Earth's crust. They contain **compounds** of metals in sufficient amounts to make it worthwhile extracting them.

Some of these compounds are **metal oxides**, for example, iron oxide (haematite). The method of extracting a metal from its ore depends on the metal's position in the **reactivity series**.

Reactivity Series

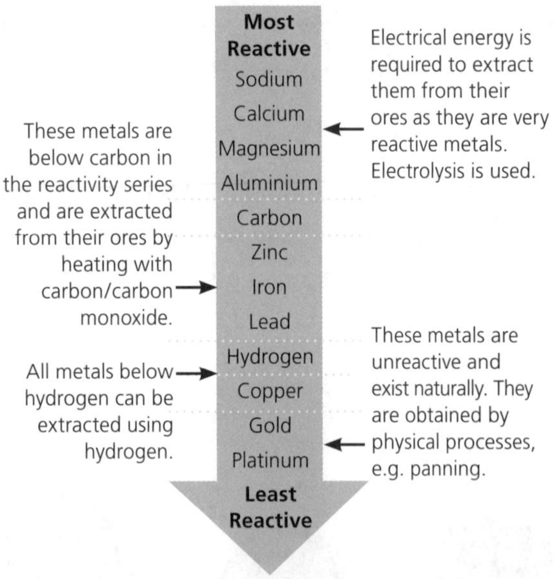
Most Reactive
Sodium
Calcium
Magnesium
Aluminium
Carbon
Zinc
Iron
Lead
Hydrogen
Copper
Gold
Platinum
Least Reactive
Electrical energy is required to extract them from their ores as they are very reactive metals. Electrolysis is used.
These metals are below carbon in the reactivity series and are extracted from their ores by heating with carbon/carbon monoxide.
All metals below hydrogen can be extracted using hydrogen.
These metals are unreactive and exist naturally. They are obtained by physical processes, e.g. panning.

The most reactive metals form the most stable ores and are therefore most difficult to extract.

The least reactive metals are found uncombined in the Earth's crust and are the easiest to extract from their ores.

Electrolysis of Molten Aluminium Oxide

Aluminium is extracted from bauxite (impure aluminium oxide). It takes a lot of energy to melt bauxite so it is first mixed with another ore of aluminium called **cryolite** as this will reduce the melting point of the bauxite. Bauxite has to be molten before it can be electrolysed to produce aluminium.

During this process:

Aluminium oxide → Aluminium + Oxygen

HT $$2Al_2O_{3(l)} \longrightarrow 4Al_{(l)} + 3O_{2(g)}$$

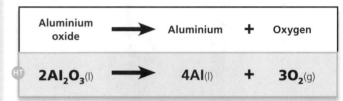

HT At the positive electrode

$$6O^{2-}_{(l)} \xrightarrow{\text{Oxidation}} 3O_{2(g)} + 12e^-$$

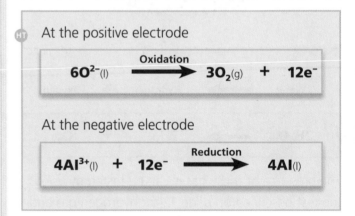

At the negative electrode

$$4Al^{3+}_{(l)} + 12e^- \xrightarrow{\text{Reduction}} 4Al_{(l)}$$

Extraction of Iron

To extract iron from its ore, **haematite** (iron(III) oxide) is smelted with carbon (in the form of coke) and limestone. (Limestone is only used to get rid of the waste.) This is a combination of reduction and oxidation processes (see page 15) that takes place inside the **blast furnace**.

The iron oxide is reduced to iron by carbon monoxide gas that is produced by the smelting processes.

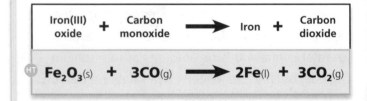

Iron(III) oxide + Carbon monoxide → Iron + Carbon dioxide

HT $$Fe_2O_{3(s)} + 3CO_{(g)} \longrightarrow 2Fe_{(l)} + 3CO_{2(g)}$$

Reduction

Reduction is the **loss of oxygen** from a compound during a chemical reaction. It is the process through which a metal compound is broken down to give the metal element. For example:

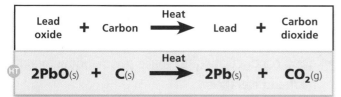

Lead oxide	+	Carbon	$\xrightarrow{\text{Heat}}$	Lead	+	Carbon dioxide
HT $2PbO_{(s)}$	+	$C_{(s)}$	$\xrightarrow{\text{Heat}}$	$2Pb_{(s)}$	+	$CO_{2(g)}$

Demonstrating Reduction of a Metal Oxide

By mixing together a spatula of carbon and copper oxide and then heating them strongly, a pinkish brown powder of copper will be formed.

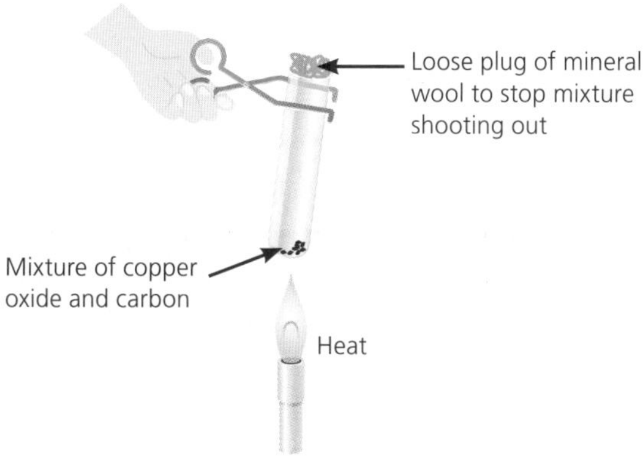

Loose plug of mineral wool to stop mixture shooting out
Mixture of copper oxide and carbon
Heat

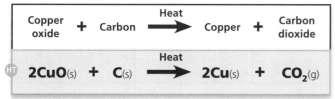

Copper oxide	+	Carbon	$\xrightarrow{\text{Heat}}$	Copper	+	Carbon dioxide
HT $2CuO_{(s)}$	+	$C_{(s)}$	$\xrightarrow{\text{Heat}}$	$2Cu_{(s)}$	+	$CO_{2(g)}$

Oxidation

When a metal or any other element combines with oxygen to make another substance, it is called **oxidation**. Oxidation is the **gaining of oxygen** by an element. For example, when magnesium is heated with oxygen it is oxidised to produce magnesium oxide.

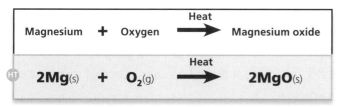

Magnesium	+	Oxygen	$\xrightarrow{\text{Heat}}$	Magnesium oxide
HT $2Mg_{(s)}$	+	$O_{2(g)}$	$\xrightarrow{\text{Heat}}$	$2MgO_{(s)}$

Corrosion of Metals

Metals are able to corrode. **Corrosion** happens when the metal reacts with oxygen in the air. The metal becomes oxidised because it has combined with oxygen to form the metal oxide.

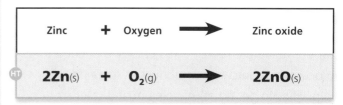

Zinc	+	Oxygen	$\longrightarrow$	Zinc oxide
HT $2Zn_{(s)}$	+	$O_{2(g)}$	$\longrightarrow$	$2ZnO_{(s)}$

Reactivity Series

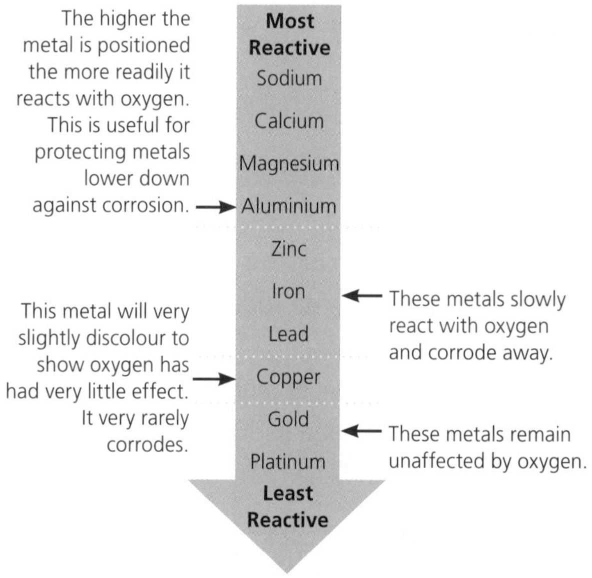

Most Reactive
Sodium
Calcium
Magnesium
Aluminium
Zinc
Iron
Lead
Copper
Gold
Platinum
Least Reactive
The higher the metal is positioned the more readily it reacts with oxygen. This is useful for protecting metals lower down against corrosion.
This metal will very slightly discolour to show oxygen has had very little effect. It very rarely corrodes.
These metals slowly react with oxygen and corrode away.
These metals remain unaffected by oxygen.

Corroded Car Made of Iron

Uncorroded Car Made of Aluminium

Recycling Metals

When we **recycle** something that we no longer need, we make a new product from the old material and at the same time we are able to conserve our planet's resources.

More and more materials can be recycled but the most common ones are metals, paper and glass.

When metals are recycled less energy is needed to process them into a new product as there are no impurities to get rid of. Aluminium and steel are currently the easiest metals to recycle.

Economic and Environmental Considerations

Recycling makes environmental sense because it:

- saves on raw materials
- saves on landfill sites and costs associated with waste disposal
- needs less energy to process the material than when the substance was first produced, so it saves on fossil fuels
- cuts down on excavation and mining so there is less environmental damage and fewer waste products
- uses less water and chemicals so reduces pollution.

Industries will tend to invest in recycling if it saves them money.

Some resources can be sustained over a long period by managing the resource, e.g. planting a tree for every tree that is cut down.

Sustainable development involves helping people to satisfy their basic requirements and enjoy a good standard of living without compromising future generations.

Properties and Uses of Metals

Aluminium, copper and gold are examples of **pure** metals. A pure metal contains atoms of that element only. Most metals extracted from the ground need to be purified.

When metals have been purified they will show many of the following general properties:

- good conductivity (they make good conductors of heat and electricity)
- dense (feel heavy)
- malleable (can be hammered into shape without cracking)
- shiny when polished
- high melting points
- strong under tension and compression
- sonorous (will ring when hit)
- ductile (can be drawn into a wire).

The following table shows the properties and uses of three pure metals.

Metal	Uses	Property
Aluminium	High-voltage power cables, furniture, drinks cans, foil food wrap	Corrosion resistant, ductile, malleable, good conductivity, low density
Copper	Electrical wiring, water pipes, saucepans	Ductile, malleable, good conductivity
Gold	Jewellery, electrical junctions	Ductile, shiny, good conductivity

Metals can also be alloyed. An **alloy** is a mixture of metals, usually produced to make the original metal stronger or improve its resistance to corrosion.

Different metals have atoms of different sizes. In an alloy, the mixture of sizes makes it harder for layers of metal ions to slide over each other, so the alloy is stronger than the individual pure metals. In a similar way, **conglomerate** rock is stronger than sandstone. Conglomerate contains grains and pebbles of various sizes, whereas sandstone contains grains of roughly the same size.

Alloys

Many metals are far more useful when they are **not pure**. Mixing a metal with other metals can change the properties of the metal.

The resulting alloy has a greater range of uses than the original metal. Remember, an alloy is a **mixture**. An alloy can have:

- a lower melting point (useful for solder)
- increased corrosion resistance (useful for anything that will be exposed to air and water)
- increased chemical resistance (useful for storing chemicals)
- increased strength and hardness (useful in construction of bridges, aircraft, cars, etc.).

AL56 WXV

Examples of Using an Alloy

Pure iron is not good for building things because it is too soft and bends easily. It also corrodes easily. If a small amount of carbon is present in iron, mild steel is produced, which is hard and strong, so it can be used for building things.

Steel is an example of an alloy of iron.

Mild Steel	99.5% Fe, 0.5% C	Hard but easily worked
Hard Steel	99% Fe, 1% C	Very hard so higher strength but brittle
Duriron	84% Fe, 1% C, 15% Si	Not affected by acid

If nickel and chromium are mixed with iron, stainless steel is produced, which is hard and rustproof, so can be used in areas that are exposed to air and water, e.g. for cutlery.

HT Nitinol

Nitinol is a mixture of nickel and titanium. It is also an example of a smart or shape memory alloy because it will return to its original form after any stress has been released. For example, nitinol will return to its original shape after it has been heated in hot water or an electrical current has been passed through it. It is mostly used in medical applications, such as spectacle frames and stents (supports) in broken or damaged blood vessels.

Pure Gold

Pure gold is a very soft metal. Gold is too soft for many applications because it can be damaged easily. To make it stronger, gold is alloyed with other metals. For yellow gold applications, it is alloyed with copper, silver and zinc. For white gold applications, it is alloyed with nickel, copper and zinc or palladium, copper and silver.

The purity of gold is measured in carats and fineness. A carat is one way of showing the proportion of gold present in an alloy, on a scale of 1 to 24. For example, 100% pure gold would be classified as 24 carat, an alloy containing 75% pure gold would be classified as 18 carat and an alloy containing 50% pure gold would be classified as 12 carats.

Fineness is measured with a three-figure numbering system. This is also seen in hallmarking of gold items. Purest gold is marked as 999 and 75% gold is marked as 750.

C1 Topic 5: Fuels

This topic looks at:

- what crude oil is and what it can be separated into
- how properties of fractions differ
- complete and incomplete combustion and the associated dangers
- the advantages and disadvantages of biofuels and hydrogen cells
- how some polymers are made, and their everyday uses and problems
- equations of polymerisation reactions

Crude Oil and Hydrocarbons

Crude oil is a mixture of **hydrocarbons** (i.e. compounds containing only hydrogen and carbon). The properties of the different hydrocarbons in crude oil remain unchanged and specific. This makes it possible to separate them by **fractional distillation**.

The oil is evaporated by heating, then allowed to condense at a range of different temperatures to form fractions. Each resulting fraction contains hydrocarbon molecules with a similar number of carbon atoms. The process takes place in a **fractionating column**.

As hydrocarbons get larger (i.e. the greater the number of carbon atoms in a molecule) they become:

- more viscous
- less flammable
- less volatile

and they have a higher boiling point.

Hydrocarbon	No. of Carbon Atoms
Refinery gases	1–4
Gasoline	5–6
Kerosene	10–16
Diesel oil	15–22
Fuel oil	30–40
Bitumen	50+

Fractional Distillation

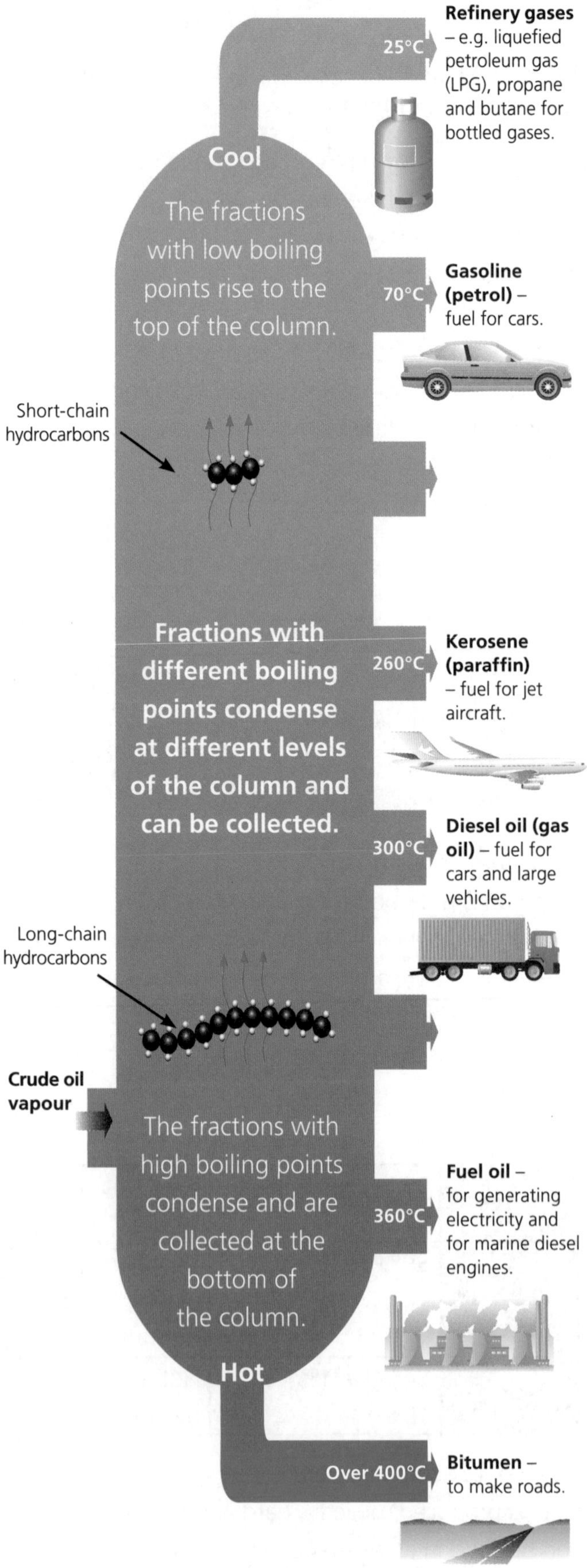

25°C
Refinery gases – e.g. liquefied petroleum gas (LPG), propane and butane for bottled gases.
Cool
The fractions with low boiling points rise to the top of the column.
70°C
Gasoline (petrol) – fuel for cars.
Short-chain hydrocarbons
Fractions with different boiling points condense at different levels of the column and can be collected.
260°C
Kerosene (paraffin) – fuel for jet aircraft.
300°C
Diesel oil (gas oil) – fuel for cars and large vehicles.
Long-chain hydrocarbons
Crude oil vapour
The fractions with high boiling points condense and are collected at the bottom of the column.
360°C
Fuel oil – for generating electricity and for marine diesel engines.
Hot
Over 400°C
Bitumen – to make roads.

Combustion

A **fuel** is a substance that releases useful amounts of energy when burned. Many fuels are hydrocarbons, e.g. methane from natural gas.

When a fuel burns it reacts with oxygen from the air.

Fuels burn with different flame colours, e.g. coal burns with a dirty yellow flame and methylated spirits burn with a purple flame.

Complete Combustion

When a hydrocarbon burns and there is plenty of oxygen available, complete combustion occurs, producing carbon dioxide (which can be can tested for by bubbling through limewater) and water, and releasing **energy**.

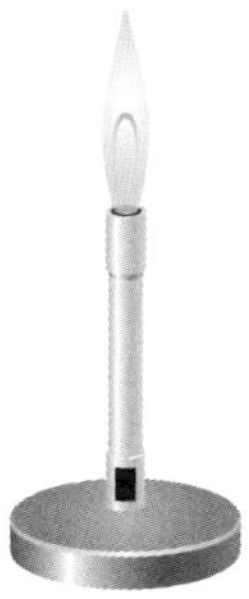

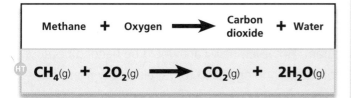

Methane + Oxygen → Carbon dioxide + Water

HT $CH_4(g) + 2O_2(g) \longrightarrow CO_2(g) + 2H_2O(g)$

Incomplete Combustion

Sometimes a fuel burns without sufficient oxygen, e.g. in a room with poor ventilation. Then, incomplete combustion takes place. Instead of carbon dioxide being produced, carbon monoxide is formed.

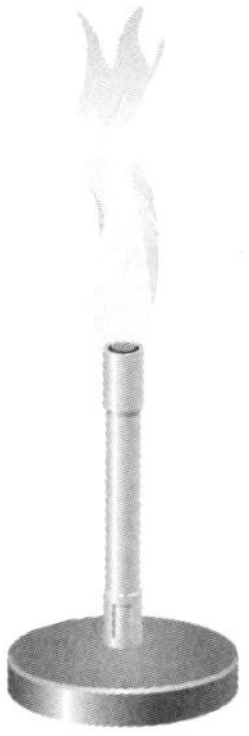

Methane + Oxygen → Carbon monoxide + Water

HT $2CH_4(g) + 3O_2(g) \longrightarrow 2CO(g) + 4H_2O(g)$

Incomplete combustion producing carbon monoxide can occur in faulty gas appliances and other heating appliances. This can be dangerous as carbon monoxide is a toxic gas.

If there is *very* little oxygen available when a hydrocarbon burns, carbon is produced. A sooty, yellow flame is an indication of incomplete combustion because it contains carbon.

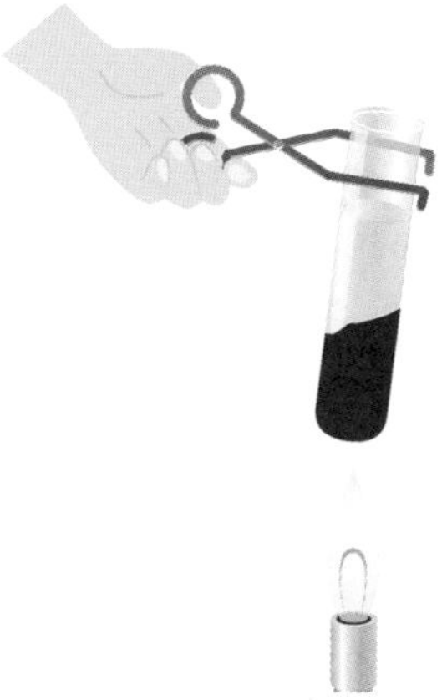

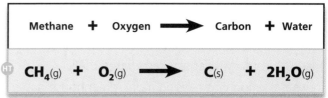

Methane + Oxygen → Carbon + Water

HT $CH_4(g) + O_2(g) \longrightarrow C(s) + 2H_2O(g)$

Although hydrocarbons produce useful amounts of energy when they burn, the gases they produce are pollutants.

- Carbon dioxide contributes to an increase in the levels of greenhouse gases, resulting in the greenhouse effect.
- Carbon monoxide, a toxic, colourless and odourless gas, combines irreversibly with the haemoglobin in red blood cells, reducing the oxygen-carrying capacity of the blood. (This eventually results in death through a lack of oxygen reaching body tissues.)

Acid Rain

Hydrocarbon fuels such as petrol, kerosene and diesel oil from crude oil, and methane from natural gas are all **fossil fuels**. Fossil fuels are non-renewable energy sources and can contain impurities such as sulfur. When fossil fuels are burned, **sulfur dioxide** is formed. Sulfur dioxide dissolves easily in rain water to form acid rain.

There are many problems associated with acid rain, which include damage to plants, animals, buildings and metals, and an increase in the acidity levels of many lakes and rivers. As acid rain is formed from gases that dissolve within the moisture in the clouds, it can fall as acid rain many miles away from where the sulfur dioxide was originally released.

Global Warming

Global warming is an accelerated increase in the average temperature of the surface of the Earth.

Much of the heat that reaches us from the Sun is reflected back from the Earth's surface. However, carbon dioxide, methane and water vapour in the atmosphere absorb heat and radiate it back down to Earth. This means that the surface of the Earth is warmer than it would be without these gases.

Scientific evidence shows that the levels of carbon dioxide in the atmosphere have varied over time. In the 20th century, scientists recognised that industrialisation was causing more 'greenhouse gases', including carbon dioxide, to remain inside the Earth's atmosphere (see also page 6).

Throughout the Earth's history, its temperature has fluctuated. Scientists are able to measure the proportion of carbon dioxide in the atmosphere and changes to global temperature over numbers of decades. Using this data, they have found a correlation between the proportion of carbon dioxide in the atmosphere and the global temperature, providing scientists with evidence for climate change.

Chemists have now developed technologies that allow them to remove carbon dioxide from the atmosphere. They can do this by adding iron dust to the ocean which is thought to enhance the growth of plankton, which takes in carbon dioxide (and therefore reduces the amount of carbon in the atmosphere). This is called **iron-seeding**.

Chemists can also convert carbon dioxide in the atmosphere into a simple hydrocarbon, by using a catalyst.

Alternatives to Fossil Fuels

Biofuels

Biofuels are fast becoming an attractive alternative to fossil fuels. These are fuels based on sustainable resources such as wood and alcohol from plants.

For some biofuels there is a balance between the carbon dioxide removed from the atmosphere during photosynthesis and the carbon dioxide produced when the fuels are transported and burned.

Ethanol

Ethanol is an alcohol produced by the fermentation of sugar beet and sugar cane. It can be used as a fuel for vehicles in its pure form, but is often added to petrol to reduce the overall need for petrol and to improve vehicle emissions. However, large areas of fertile land are needed to grow the crops.

Ethanol + Oxygen → Carbon dioxide + Water

HT $$C_2H_5OH_{(l)} + 3O_{2(g)} \longrightarrow 2CO_{2(g)} + 3H_2O_{(g)}$$

A fuel cell is a way of producing and harnessing energy from a chemical reaction. A hydrogen fuel cell uses hydrogen, which reacts with oxygen, usually from the air. **Hydrogen** is the cleanest of all the fuels as it only produces water. If used in cars, it can supply three times the energy of petrol per gram. However, new cars would be needed because of the technology used to get electricity from the hydrogen. Producing the hydrogen involves the use of fossil fuels so still adds to pollution.

The chemical reaction in a simple fuel cell is:

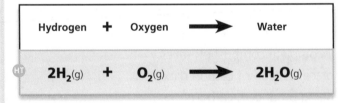

Hydrogen + Oxygen → Water

HT $$2H_{2(g)} + O_{2(g)} \longrightarrow 2H_2O_{(g)}$$

Good Fuels

There are many factors to be considered when deciding if a fuel is a good fuel, including these examples.

- Is the fuel safe in that it does not catch alight too easily, especially when transported?
- Is the fuel easy to transport? Liquid fuels are far easier to transport and store, compared to gases.
- Is the fuel a good provider of heat energy so that not too much fuel has to be burned for the energy generated?

- Is it is a clean fuel that produces very little particulate or toxic products, or pollutants that contribute to greenhouse gases?

Different fuels release different amounts of energy when burned. Fuels can be compared by measuring the temperature rise of a fixed volume of water as it is heated by a known amount of fuel.

Alkanes (Saturated Hydrocarbons)

An alkane is a hydrocarbon in which each carbon atom is bonded to four other atoms, each of which is either a carbon or a hydrogen atom. When more than one carbon atom is present they are joined by a **single covalent carbon–carbon bond** to form a chain or 'spine' that runs the length of the molecule. We say that alkanes are **saturated** because each carbon atom is bonded to the maximum number of atoms.

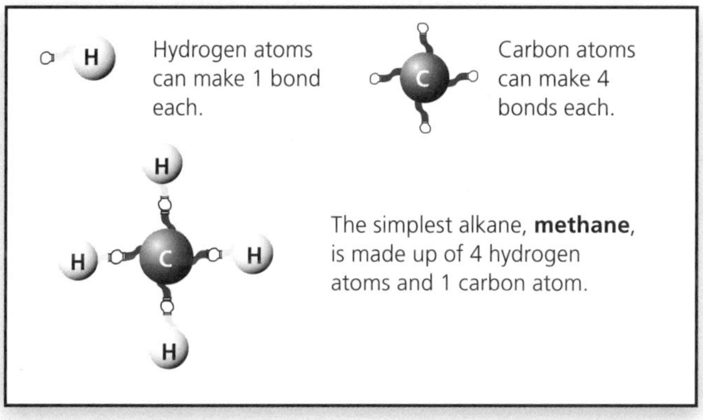
H
Hydrogen atoms can make 1 bond each.
C
Carbon atoms can make 4 bonds each.
The simplest alkane, **methane**, is made up of 4 hydrogen atoms and 1 carbon atom.

The next simplest alkanes are **ethane** and **propane**. Because they are saturated (i.e. the carbon atoms are all bonded to four other atoms), they are fairly unreactive, although they do burn well.

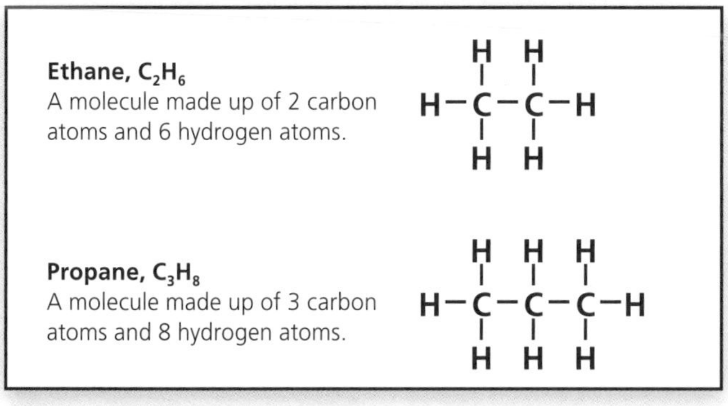
Ethane, C_2H_6
A molecule made up of 2 carbon atoms and 6 hydrogen atoms.
Propane, C_3H_8
A molecule made up of 3 carbon atoms and 8 hydrogen atoms.

Alkenes (Unsaturated Hydrocarbons)

The **alkenes** are another kind of hydrocarbon. They are very similar to the alkanes, except that two of the carbon atoms are joined by a **double covalent carbon–carbon bond**. Because the carbon atoms are not bonded to the maximum number of atoms, we call them **unsaturated hydrocarbons**.

The simplest alkene is **ethene**, C_2H_4, which is made of four hydrogen atoms and two carbon atoms. Ethene contains one double carbon–carbon bond.

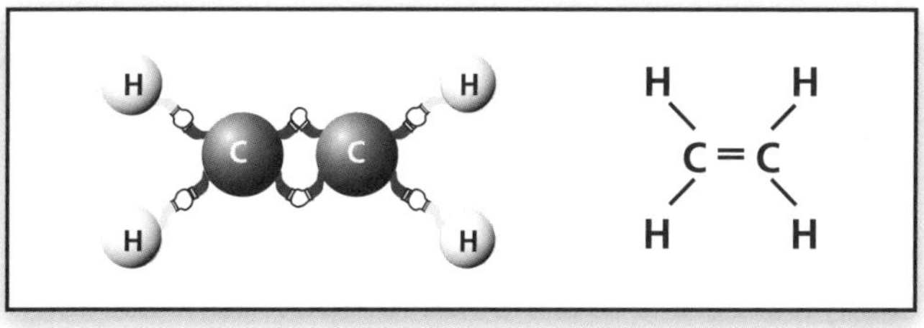

The next simplest alkene is **propene**, C_3H_6, which is made of six hydrogen atoms and three carbon atoms. Propene contains one double carbon–carbon bond and one single carbon–carbon bond.

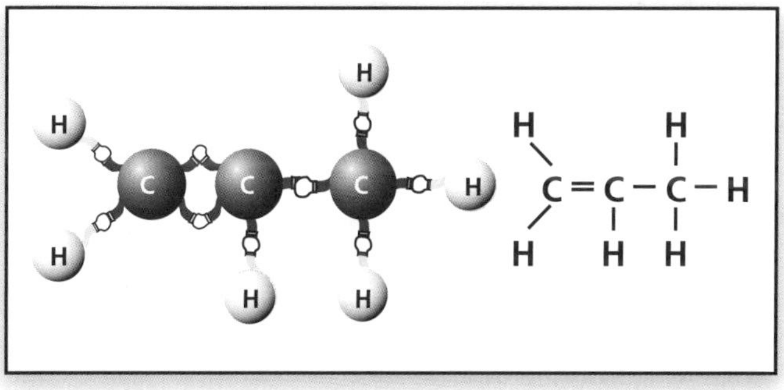

Testing for Alkanes and Alkenes

A simple test to distinguish between alkanes and alkenes is to add bromine water.

Alkenes will **decolourise** bromine water as the alkene reacts with it. Alkanes have no effect on bromine water.

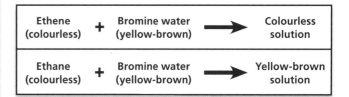

Ethene (colourless)	**+**	**Bromine water (yellow-brown)**	→	**Colourless solution**
Ethane (colourless)	**+**	**Bromine water (yellow-brown)**	→	**Yellow-brown solution**

Cracking

Cracking is the breaking down of long-chain hydrocarbon molecules into more useful short-chain hydrocarbon molecules, which are then used to help meet demands for hydrocarbons in short supply.

Long-Chain Hydrocarbon

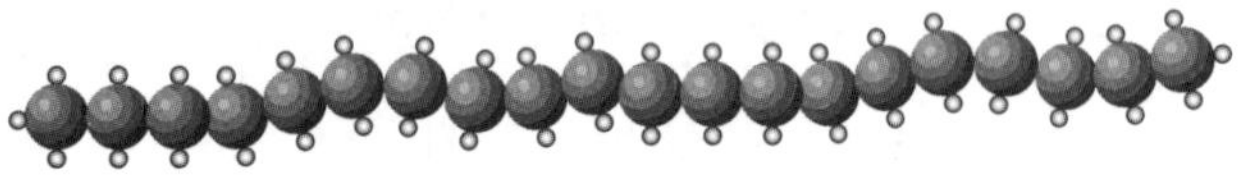

Short-Chain Hydrocarbons

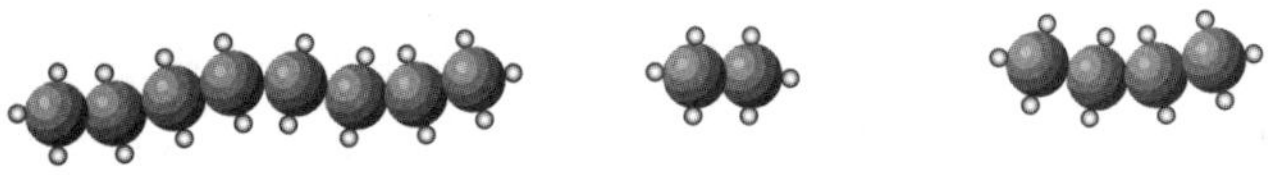

The long-chain hydrocarbon is heated until it vaporises. The vapour is then passed over a heated catalyst where a **thermal decomposition** reaction takes place.

In the laboratory, cracking can be carried out using the following apparatus:

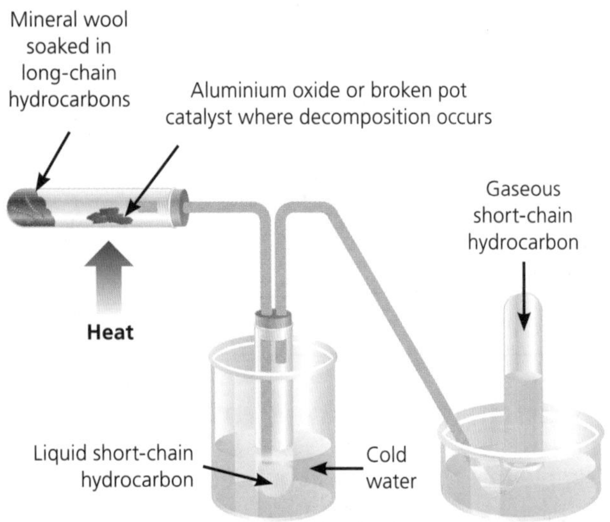
Mineral wool soaked in long-chain hydrocarbons
Aluminium oxide or broken pot catalyst where decomposition occurs
Gaseous short-chain hydrocarbon
Heat
Liquid short-chain hydrocarbon
Cold water

When alkanes are cracked, alkanes and alkenes are formed, for example:

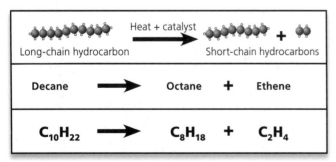

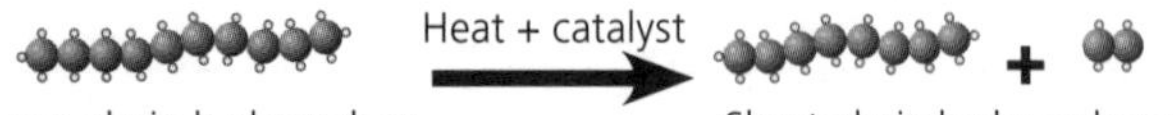 Long-chain hydrocarbon	Heat + catalyst →	Short-chain hydrocarbons	+	
Decane	→	**Octane**	**+**	**Ethene**
$C_{10}H_{22}$	→	C_8H_{18}	+	C_2H_4

HT Supply and Demand

This bar chart shows:

- the relative amounts of each fraction in crude oil
- the demand for each fraction in crude oil.

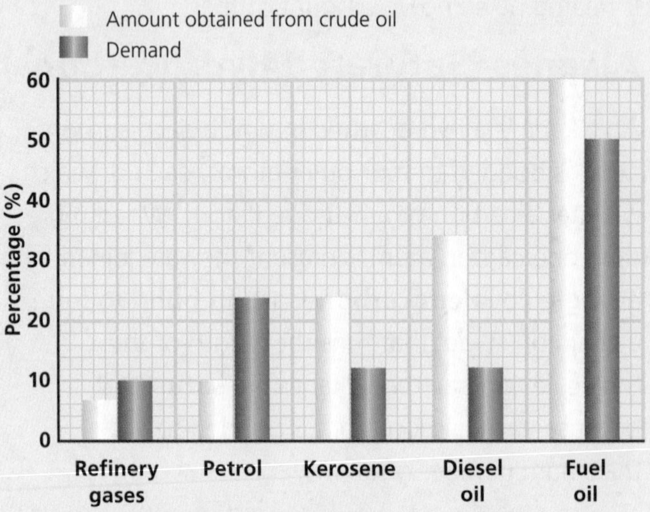
Amount obtained from crude oil
Demand
Percentage (%)
60
50
40
30
20
10
0
Refinery gases
Petrol
Kerosene
Diesel oil
Fuel oil

The demand for some fractions is **greater than** the supply, especially for shorter-chain hydrocarbons, such as petrol. This is because they release energy more quickly by burning, so they make better fuels.

Longer-chain hydrocarbons are broken down into more useful shorter-chain hydrocarbons through cracking. The compounds that are produced have a variety of uses, for example to fuel transport, to heat homes and to make chemicals for drugs.

As crude oil runs out, supply decreases, but demand remains the same and so prices increase.

Scientists are always trying to find alternative ways to make us less reliant on fossil fuels.

Polymers

Long-chain hydrocarbon molecules are often **polymers** made up from smaller **monomers**. A monomer is a short-chain unsaturated hydrocarbon molecule.

Alkenes can be good at joining together because they are unsaturated. When they join together without producing another substance, we call this **addition polymerisation**. For example ethene monomers form the polymer poly(ethene).

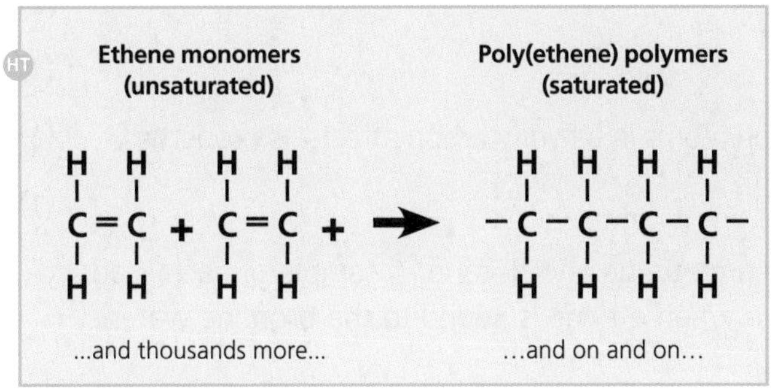

Ethene monomers (unsaturated) ...and thousands more... **Poly(ethene) polymers (saturated)** ...and on and on...

The general formula for addition polymerisation is:

$$n\left(\begin{array}{c} | \; | \\ C=C \\ | \; | \end{array}\right) \longrightarrow \left(\begin{array}{c} | \; | \\ C-C \\ | \; | \end{array}\right)_n$$

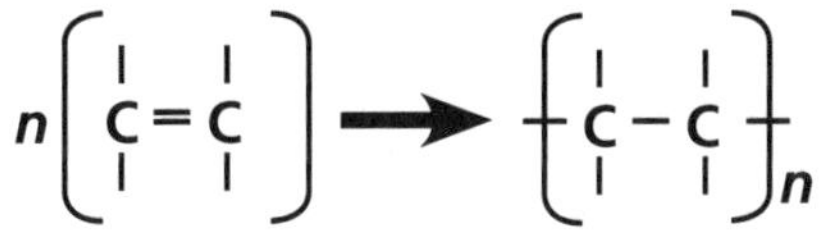

where *n* is a very large number

We can use this formula with different monomer units to show how poly(propene), poly(chloroethene) and PTFE are formed.

Poly(propene):

$$n\left(\begin{array}{cc} H & CH_3 \\ | & | \\ C=C \\ | & | \\ H & H \end{array}\right) \longrightarrow \left(\begin{array}{cc} H & CH_3 \\ | & | \\ C-C \\ | & | \\ H & H \end{array}\right)_n$$

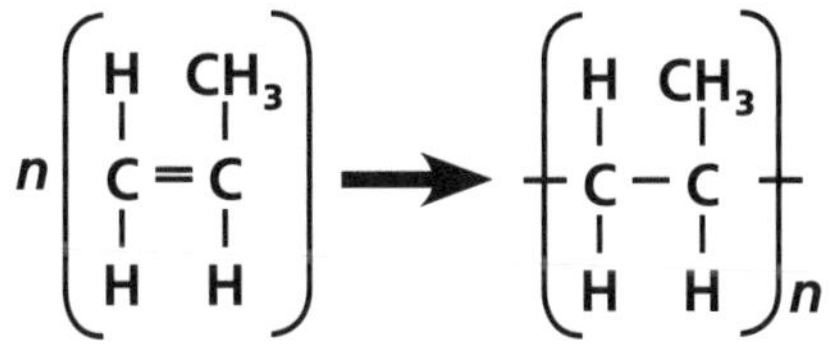

Poly(chloroethene):

$$n\left(\begin{array}{cc} H & H \\ | & | \\ C=C \\ | & | \\ H & Cl \end{array}\right) \longrightarrow \left(\begin{array}{cc} H & H \\ | & | \\ C-C \\ | & | \\ H & Cl \end{array}\right)_n$$

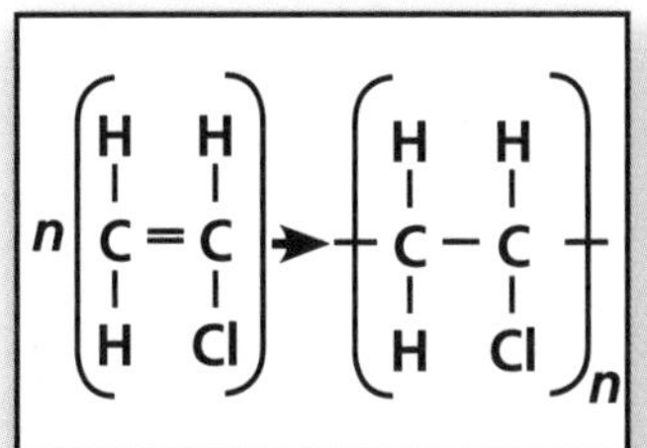

PTFE:

$$n\left(\begin{array}{cc} F & F \\ | & | \\ C=C \\ | & | \\ F & F \end{array}\right) \longrightarrow \left(\begin{array}{cc} F & F \\ | & | \\ C-C \\ | & | \\ F & F \end{array}\right)_n$$

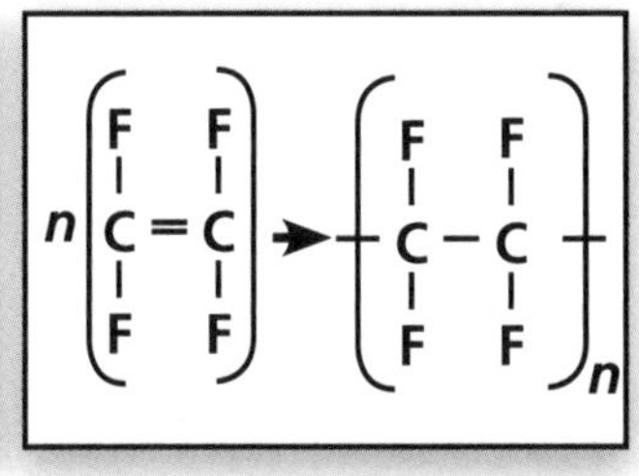

Uses of Some Polymers

Generally polymers such as poly(ethene), poly(propene), poly(chloroethene) (PVC) and PTFE are very versatile as they have a variety of different uses.

Polymer	Uses
Poly(ethene), PE Good chemical resistance, flexible	Food coverings, bins, water pipes, bags, bottles, food trays
Poly(propene), PP Good chemical resistance, flexible	Chairs, stationery, ropes, crates, bottles, clothing
Poly(chloroethene), PVC Good impact resistance, rigid or flexible	Flooring, double-glazed window frames, pipes, clothing, cable covering, electrical components
PTFE Water-resistant, flexible, non-stick	Non-stick cookware coating, lubricant sprays, to make waterproof coatings

Disposing of Polymers

There are various ways of disposing of polymers. Unfortunately, some of these methods can be harmful to the environment.

- **Burning polymers** produces air pollution. Some polymers should not be burned as they produce toxic fumes, for example burning PVC produces hydrogen chloride gas.
- **Use of landfill sites** means that plastic waste builds up because most polymers are non-biodegradable. Microorganisms have no effect on them: they will not decompose and rot away.
- **Recycling** some polymers uses less energy to make a new product, reduces the need for oil in polymer production, reduces the amount of polymer waste going to landfill and helps reduce carbon dioxide emissions.

More and more companies have made investments into biodegradable polymers, which are continually being developed.

Exam Practice Questions

Questions labelled with an asterisk (*) are ones where the quality of your written communication will be assessed – you should take particular care with your spelling, punctuation and grammar, as well as the clarity of expression, on these questions.

1 **(a)** An acid–base reaction is called:

A ☐ oxidation **B** ☐ neutralisation **C** ☐ reduction **D** ☐ precipitation **(1)**

(b) What type of product would be formed in this type of reaction? **(1)**

2 **(a)** How is the atmosphere today different from the Earth's early atmosphere? **(2)**

(b) Describe how water vapour from early volcanic activity formed the oceans. **(2)**

3 State three ways in which the levels of carbon dioxide in the Earth's atmosphere have increased over the last 100 years. **(3)**

4 **(a)** What is a hydrocarbon? **(1)**

(b) State one way in which increasing the number of carbon atoms in a hydrocarbon affects its properties. **(1)**

(c) Why can crude oil be considered as a mixture? **(1)**

(d) Two hydrocarbons obtained from a cracking process are ethene and decane. A sample of each is added to bromine water. What would you expect to happen when ethene is added to the bromine water?

A ☐ The solution turns yellow.

B ☐ The solution turns from orange to colourless.

C ☐ The solution turns milky.

D ☐ The solution remains orange. **(1)**

5 **(a)** Give one economic advantage to recycling metals. **(1)**

***(b)** Explain in as much detail as you can why recycling can be considered as sustainable. **(6)**

6 **(a)** What is an ore? **(1)**

(b) Which property of a metal determines how easily it is extracted from an ore? **(1)**

(c) **(i)** Copper oxide can be heated with carbon to extract the metal. Complete the following equation.

Copper oxide + Carbon ⟶ .... + ... **(1)**

(ii) What has happened to the copper oxide in the above equation? **(1)**

(d) What does the term 'reduction' mean? **(1)**

7 Explain why the method used to extract aluminium from its oxide is different from that used to extract iron from its oxide. **(4)**

8 **(a)** What is produced when calcium carbonate is decomposed? **(2)**

(b) Limestone is a very important building material that can be broken down by thermal decomposition. Why is limestone important in the building industry? **(2)**

(c) Describe four factors that should be taken into consideration when quarrying for limestone. Your answer must include at least one each of the environmental, economic and social effects. **(4)**

9 Acid and alkali reactions all follow the same general equation.

(a) Write the word equation for the reaction of sulfuric acid and magnesium oxide. **(1)**

(b) Write the word equation for the reaction of nitric acid and copper oxide. **(1)**

(c) Describe why ingesting indigestion remedies is the same as a neutralisation reaction. **(2)**

(d) Why does the stomach contain hydrochloric acid? **(2)**

10 **(a)** Describe the difference between a monomer and a polymer. **(2)**

(b) Give the structural formula for PTFE. **(1)**

(c) Choose the correct answer to complete the following sentence.

When monomers join together without producing another substance, it is called

A ☐ co-polymerisation. **B** ☐ addition polymerisation.

C ☐ polymerisation. **D** ☐ condensation polymerisation. **(1)**

(d) Explain why burning polymers is not an environmentally friendly way of disposing of them. **(3)**

(e) State one property of PTFE that makes it useful as a lubricant. **(1)**

11 Chlorine is a very important element. It can be obtained from the electrolysis of sea water and from brine, which is the basis of the chlor-alkali industry.

(a) Describe how electrolysis breaks down brine to release chlorine gas and hydrogen gas. **(3)**

(b) (i) Describe the appearance of chlorine gas. **(1)**

(ii) How would you test for chlorine gas at the positive terminal of an electrolysis cell? **(1)**

(c) Why is the large-scale manufacture of chlorine considered to be potentially hazardous? **(2)**

12 *Iron is a very versatile metal. It is extracted from haematite, iron(III) oxide, by reacting it with carbon monoxide in a blast furnace. Pure iron has very few uses as it is too soft. To make it more useful iron is mixed with other elements.

Explain in as much detail as you can how mixing iron with other elements improves its properties. **(6)**

HT 13 This question is about fuels.

(a) (i) Methane has the formula CH_4 and is the fuel that is often called 'natural gas'. It is also an alkane. Write a balanced equation for complete combustion of methane in oxygen. **(1)**

(ii) When methane reacts with insufficient oxygen, incomplete combustion takes place. Write a word equation and a balanced symbol equation for this reaction. **(2)**

***(b)** Petrol and diesel oil are fractions of crude oil. As a fuel, petrol is in greater demand (25% of demand for fuels) than can be supplied (10% obtained from crude oil). This is not the case for diesel oil (12% of demand for fuels, but 34% obtained from crude oil).

Describe as fully as you can what you understand by the supply and demand of crude oil, using the examples given, and how crude oil refinery overcomes this problem. **(6)**

C2 Topic 1: Atomic Structure and the Periodic Table

This topic looks at:
- how Mendeleev developed the early periodic table
- the general trends and properties of the periodic table
- what subatomic particles are
- what isotopes and relative atomic masses are
- the electronic configurations of the first 20 elements

The History of the Periodic Table

All things are made of elements. The known elements are arranged in the **periodic table**. However, the periodic table is a relatively recent invention.
- Before they knew about chemical elements, some early scientists described things as being made from fire, air, earth and water.
- The idea of elements was first mentioned in the mid-1600s.

Before the modern periodic table was developed, a number of attempts were made to arrange the elements. However, it was not until 1869 that Russian chemist **Dimitri Mendeleev** developed the modern periodic table.
- He arranged the elements in order of atomic mass.
- The table included all the known elements and left gaps for those not yet discovered.
- Each element was put into the group (column) where its properties fitted best.
- Elements that did not fit were put into a spare column.

Using his periodic table, Mendeleev was able to predict the existence and properties of undiscovered elements. One of these was silicon, which was discovered many years later and matched the predictions.

The modern periodic table can also be used to predict the properties of artificial elements (those that do not occur naturally).

The Periodic Table

Elements are the building blocks of all materials. The 100 or so elements are arranged in the periodic table, in order of increasing atomic number. The elements are arranged in rows (**periods**) so that elements with similar properties are in the same column (**group**). This forms the basis of the periodic table (a detailed version is on page 89).

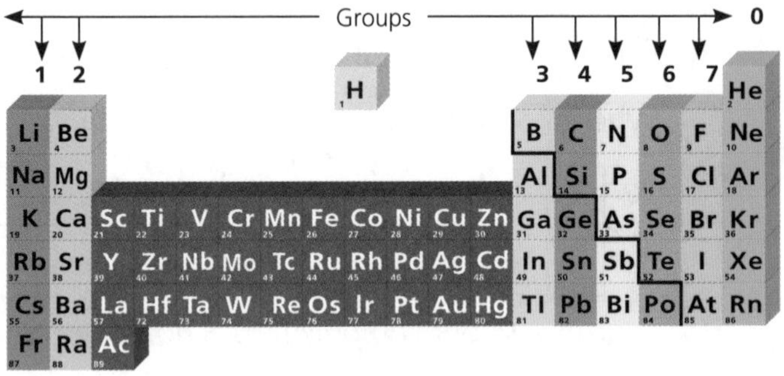

Groups
1
2
3
4
5
6
7
0
H
He
Li Be
B C N O F Ne
Na Mg
Al Si P S Cl Ar
K Ca Sc Ti V Cr Mn Fe Co Ni Cu Zn Ga Ge As Se Br Kr
Rb Sr Y Zr Nb Mo Tc Ru Rh Pd Ag Cd In Sn Sb Te I Xe
Cs Ba La Hf Ta W Re Os Ir Pt Au Hg Tl Pb Bi Po At Rn
Fr Ra Ac

- More than three-quarters of the elements are metals; the rest are non-metals.
- Metals are found mainly in Groups 1 and 2 and in the central block.
- Group 1 elements are known as the **alkali metals**.
- Group 7 elements are known as the **halogens**.
- Group 0 elements are known as the **noble gases**.
- The **transition metals** are in the central block between Group 2 and Group 3.

Trends in the Periodic Table

- The elements in a particular group have similar chemical properties since they have the same number of electrons in their outermost shells.
- The mass of elements gets bigger as you go from left to right across a period (row).
- Elements in Group 1 (which all have 1 electron in their outermost shell) become more reactive as you go down the group.
- Elements in Group 7 (which all have 7 electrons in their outermost shell) become less reactive as you go down the group.
- Elements in Group 0 all have a full outermost electron shell. All these elements are unreactive.

The Atom

Elements are made up of **atoms**. An atom has a nucleus that contains the subatomic particles **protons** and **neutrons** (the exception is hydrogen, which does not contain neutrons). The nucleus is surrounded by orbiting electrons arranged in shells. These particles have different relative masses and charges.

Subatomic Particle	Relative Mass	Relative Charge
Proton	1	+1 (positive)
Neutron	1	0 (neutral)
Electron	Negligible	–1 (negative)

An atom has the same number of protons as electrons, so the atom as a whole has no electrical charge. All the atoms of a particular element have:

- the same number of protons in their nuclei
- the same number of electrons orbiting the nucleus.

A Representation of a Helium Atom

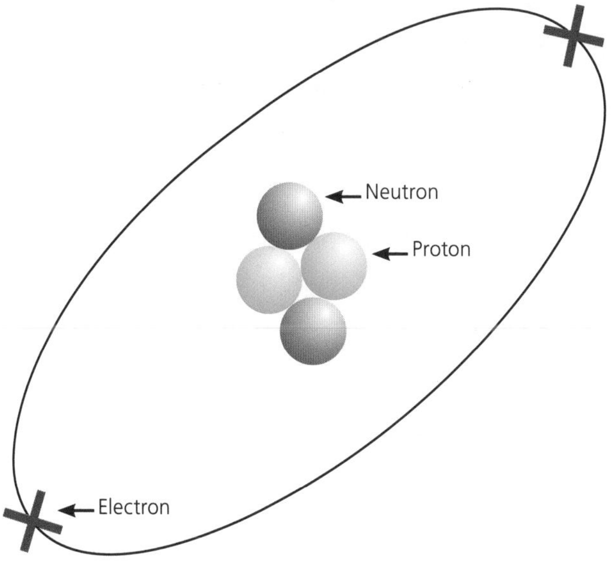
Neutron
Proton
Electron

Atomic Number and Mass Number

Each element in the periodic table has two numbers next to its symbol, for example:

Lithium $^{7}_{3}Li$ Nitrogen $^{14}_{7}N$ Iron $^{56}_{26}Fe$

The number at the bottom is the **atomic number**. This gives the number of protons found in the nucleus of an atom of that element. It also gives the number of electrons orbiting the nucleus, because all atoms have no overall electrical charge.

The number at the top is the **mass number**.

The mass number tells you the total number of protons and neutrons there are in the nucleus of an atom of the element.

Element	Atomic Number	Mass Number
Lithium (nucleus contains 3 protons and 4 neutrons)	3	7
Nitrogen (nucleus contains 7 protons and 7 neutrons)	7	14
Iron (nucleus contains 26 protons and 30 neutrons)	26	56

Relative Atomic Mass

The actual mass of a single atom is far too small to be used easily in calculations. To make things more manageable we use **relative atomic mass**.

Relative atomic mass is the mass of a particular atom compared to $\frac{1}{12}$ of the mass of a carbon atom.

Isotopes

HT

The number of **protons** in the atom defines the element. All the atoms of a particular element have the same number of protons in their nuclei; this number is unique to each particular element and is its atomic number.

However, atoms of the same element can have different numbers of **neutrons**: these atoms are called **isotopes** of the element.

Because all isotopes of an element have the same number of protons and the same electronic configuration, they have the same chemical properties and so their reactions are the same. They are easy to spot because they have the **same atomic number** but a **different mass number**.

Examples

1. Chlorine has two isotopes.

$^{35}_{17}Cl$	$^{37}_{17}Cl$
17 protons	17 protons
17 electrons	17 electrons
18 neutrons (35 – 17 = 18)	20 neutrons (37 – 17 = 20)

2. Hydrogen has three isotopes.

$^{1}_{1}H$	$^{2}_{1}H$	$^{3}_{1}H$
1 proton	1 proton	1 proton
1 electron	1 electron	1 electron
0 neutrons (1 – 1 = 0)	1 neutron (2 – 1 = 1)	2 neutrons (3 – 1 = 2)

3. Carbon has three naturally occurring isotopes.

$^{12}_{6}C$	$^{13}_{6}C$	$^{14}_{6}C$
6 protons	6 protons	6 protons
6 electrons	6 electrons	6 electrons
6 neutrons (12 – 6 = 6)	7 neutrons (13 – 6 = 7)	8 neutrons (14 – 6 = 8)

Relative Atomic Mass

The mass numbers in the examples in column 1 are the atomic masses of each particular isotope of an element. **Relative atomic mass** is a (weighted) average for the different isotopes of an element.

Chemists use relative atomic masses because they take into account the relative isotopic masses and the abundance of each one.

Example 1: Chlorine
Naturally occurring chlorine consists of 75% of $^{35}_{17}Cl$ and 25% of $^{37}_{17}Cl$, i.e. in a ratio of 3:1

So, for every four atoms of chlorine, three of them are $^{35}_{17}Cl$ and one of them is $^{37}_{17}Cl$.

So the total atomic mass of these four atoms = (3 × 35) + (1 × 37) = 142

Therefore, the **relative atomic mass** of chlorine is:

$$\frac{142}{4} = \mathbf{35.5}$$

N.B. Relative atomic masses and relative isotopic masses are often not whole numbers but are rounded up or down for display in the periodic table for ease of calculations.

Example 2: Magnesium
Magnesium consists of 80% of $^{24}_{12}Mg$, 10% of $^{25}_{12}Mg$ and 10% of $^{26}_{12}Mg$, i.e. in the ratio of 8:1:1

So, for every ten atoms of magnesium, eight of them are $^{24}_{12}Mg$, one of them is $^{25}_{12}Mg$ and one of them is $^{26}_{12}Mg$.

So the total atomic mass of these ten atoms = (8 × 24) + (1 × 25) + (1 × 26) = 243

Therefore, the **relative atomic mass** of magnesium is:

$$\frac{243}{10} = \mathbf{24.3}$$

Electronic Configuration

Electronic configuration shows how the electrons are arranged around the nucleus of an atom in energy levels (shells).

- The electrons in an atom occupy the lowest available shells (i.e. the shells nearest to the nucleus).
- The first shell can only contain a maximum of two electrons.
- The shells after the first shell can each hold a maximum of eight electrons.
- We write electronic configuration as a series of numbers.

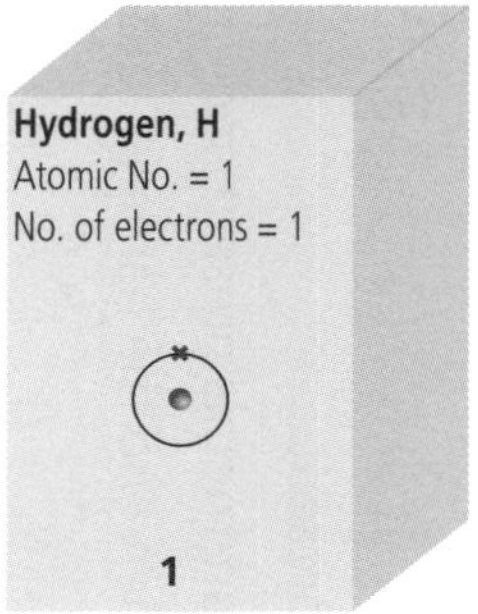
Hydrogen, H
Atomic No. = 1
No. of electrons = 1
1

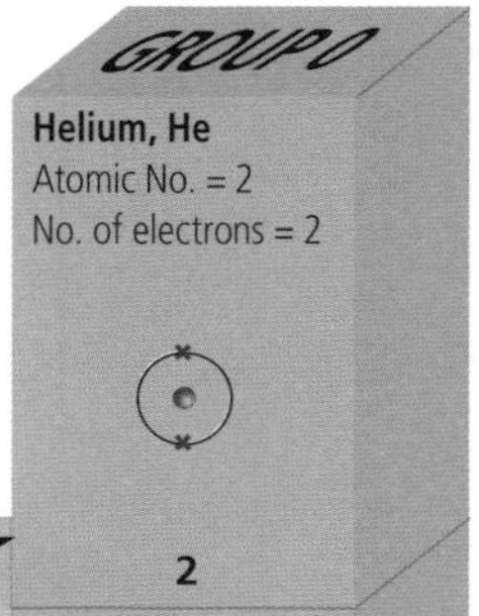
GROUP 0
Helium, He
Atomic No. = 2
No. of electrons = 2
2

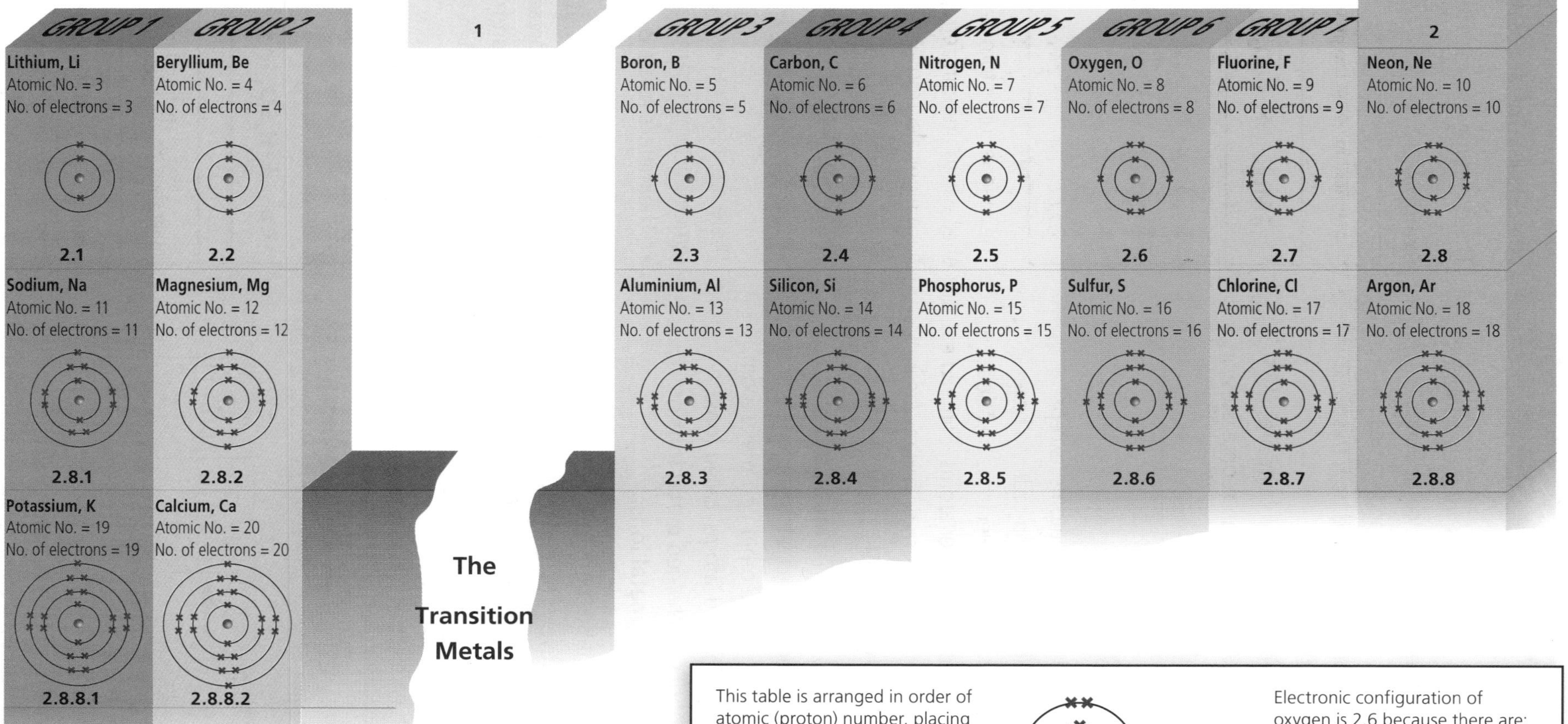
GROUP 1
GROUP 2
GROUP 3
GROUP 4
GROUP 5
GROUP 6
GROUP 7
Lithium, Li
Atomic No. = 3
No. of electrons = 3
2.1
Beryllium, Be
Atomic No. = 4
No. of electrons = 4
2.2
Boron, B
Atomic No. = 5
No. of electrons = 5
2.3
Carbon, C
Atomic No. = 6
No. of electrons = 6
2.4
Nitrogen, N
Atomic No. = 7
No. of electrons = 7
2.5
Oxygen, O
Atomic No. = 8
No. of electrons = 8
2.6
Fluorine, F
Atomic No. = 9
No. of electrons = 9
2.7
Neon, Ne
Atomic No. = 10
No. of electrons = 10
2.8
Sodium, Na
Atomic No. = 11
No. of electrons = 11
2.8.1
Magnesium, Mg
Atomic No. = 12
No. of electrons = 12
2.8.2
Aluminium, Al
Atomic No. = 13
No. of electrons = 13
2.8.3
Silicon, Si
Atomic No. = 14
No. of electrons = 14
2.8.4
Phosphorus, P
Atomic No. = 15
No. of electrons = 15
2.8.5
Sulfur, S
Atomic No. = 16
No. of electrons = 16
2.8.6
Chlorine, Cl
Atomic No. = 17
No. of electrons = 17
2.8.7
Argon, Ar
Atomic No. = 18
No. of electrons = 18
2.8.8
Potassium, K
Atomic No. = 19
No. of electrons = 19
2.8.8.1
Calcium, Ca
Atomic No. = 20
No. of electrons = 20
2.8.8.2
The Transition Metals

Notice that there is a connection between the number of outer electrons and the position of an element in a group: elements in Group 1 have only one electron in their outermost shell, elements in Group 2 have only two electrons in their outermost shell, and so on.

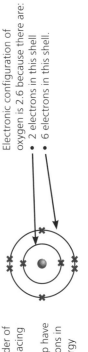

This table is arranged in order of atomic (proton) number, placing the elements in groups. Elements in the same group have the same number of electrons in their highest occupied energy level (outer shell).

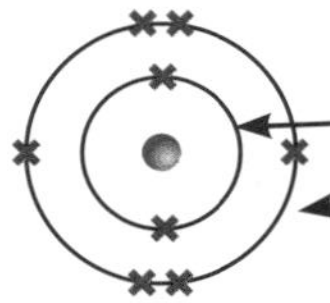

Electronic configuration of oxygen is 2.6 because there are:

- 2 electrons in this shell
- 6 electrons in this shell.

C2 Ionic Compounds and Analysis

C2 Topic 2: Ionic Compounds and Analysis

This topic looks at:

- how metals and non-metals form compounds
- the basic properties of ionic compounds
- how to work out formulae and names
- type of structure formed and how this affects properties
- what the general rules of solubility are
- how precipitates are made and some uses
- the general tests for ions

Ionic Bonding

To become chemically stable, all atoms (except hydrogen) lose or gain electrons in order to have eight in their outer shell. These electrons must be accepted by, or donated by, other atoms. Sometimes electrons are completely transferred. This results in the formation of an **ionic bond**.

An ionic bond occurs between a **metal atom** and a **non-metal atom** and involves a transfer of electrons from one metal atom to the other non-metal atom, to form electrically charged 'atoms' called **ions**, which may be positively charged **metal ions** or negatively charged **non-metal ions**.

Groups of atoms can also form ions. For example, nitrate ions are negatively charged (NO_3^-) and ammonium ions are positively charged (NH_4^+).

Example 1

Sodium	**+**	**Chlorine**	→	**Sodium chloride**
2Na(s)	**+**	Cl_2(g)	→	**2NaCl**(s)

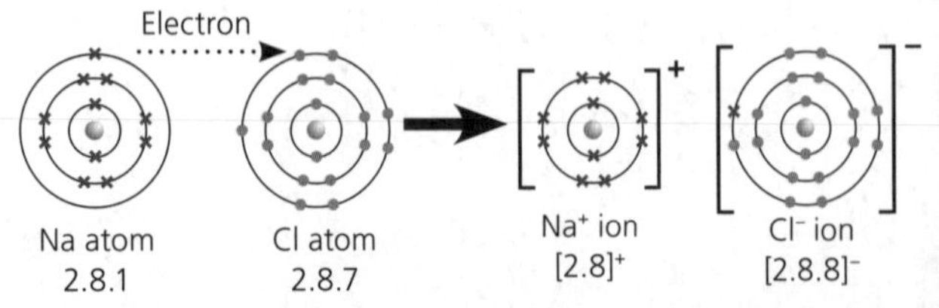

Electron
+
−
Na atom
2.8.1
Cl atom
2.8.7
Na⁺ ion
[2.8]⁺
Cl⁻ ion
[2.8.8]⁻

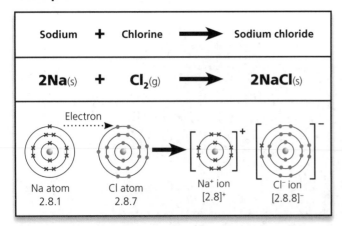

The sodium atom has one electron in its outer shell, which is transferred to the chlorine atom. Both now have eight electrons in their outer shell (i.e. complete outer shells). The atoms are now ions: Na^+ and Cl^-.

The compound formed is sodium chloride, NaCl.

Example 2

Magnesium	**+**	**Oxygen**	→	**Magnesium oxide**
2Mg(s)	**+**	O_2(g)	→	**2MgO**(s)

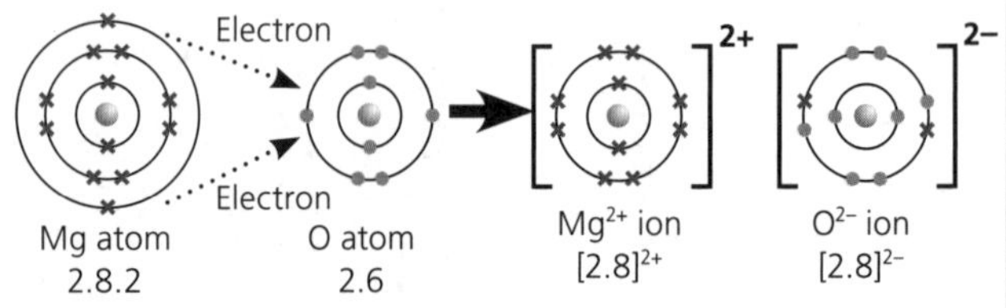

Electron
Electron
2+
2−
Mg atom
2.8.2
O atom
2.6
Mg²⁺ ion
[2.8]²⁺
O²⁻ ion
[2.8]²⁻

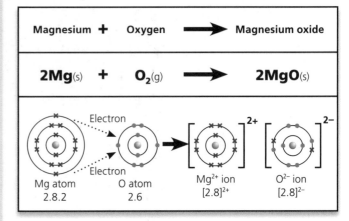

The magnesium atom has two electrons in its outer shell which are transferred to the oxygen atom. Both have eight electrons in their outer shell. The atoms are now ions: Mg^{2+} and O^{2-}.

The compound formed is magnesium oxide, MgO.

Naming Metal Compounds

There are a few general and simple rules to remember when naming metal compounds:

- The metal's name is always written first.
- Change the ending of the name of the non-metal to -ide, but only when there is a single non-metal present.
- If there are two non-metals present and one of them is oxygen, then end the name in -ate.

For example:

Calcium	**+**	**Sulfuric acid**	→	**Calcium sulfate**	**+**	**Hydrogen**
Ca(s)	**+**	H_2SO_4(aq)	→	$CaSO_4$(aq)	**+**	H_2(g)

Magesium	**+**	**Hydrochloric acid**	→	**Magnesium chloride**	**+**	**Hydrogen**
Mg(S)	**+**	**2HCl**(aq)	→	$MgCl_2$(aq)	**+**	H_2(g)

Predicting Formulae

Ionic compounds are neutral because the charges on the ions cancel each other out. Knowing this and the charge on the ions, we can predict formulae for any ionic compound.

Predicting the Formula for Magnesium Chloride
Magnesium ions have a 2+ charge, Mg^{2+}.
Chloride ions have a 1− charge, Cl^-.
The charge on two chloride ions balances out the charge on the magnesium ion: $Mg^{2+} + 2 \times Cl^- = MgCl_2$.
This is the same for any magnesium halide.

Predicting the Formula for Sodium Carbonate
Sodium ions have a 1+ charge, Na^+.
Carbonate ions have a 2- charge, CO_3^{2-}.
The charge on two sodium ions balances out the charge on the carbonate ion: $2 \times Na^+ + CO_3^{2-} = Na_2CO_3$.

Predicting the Formula for Calcium Hydroxide
Calcium ions have a 2+ charge, Ca^{2+}.
Hydroxide ions have a 1- charge, OH^-.
The charge on two hydroxide ions balances out the charge on one calcium ion: $Ca^{2+} + 2 \times OH^- = Ca(OH)_2$.
Brackets ( ) are used to show there are two hydroxide ions in the compound.

Predicting the Formula for Potassium Nitrate
Potassium ions have a 1+ charge, K^+.
Nitrate ions have a 1- charge, NO_3^-.
The charge on one nitrate ion balances out the charge on one potassium ion: $K^+ + NO_3^- = KNO_3$.

Predicting the Formula for Calcium Oxide
Calcium ions have a 2+ charge, Ca^{2+}.
Oxide ions have a 2- charge, O^{2-}.
The charge on one oxide ion balances out the charge on one calcium ion: $Ca^{2+} + O^{2-} = CaO$.

Predicting the Formula for Magnesium Sulfate
Magnesium ions have a 2+ charge, Mg^{2+}.
Sulfate ions have a 2- charge, SO_4^{2-}.
The charge on one sulfate ion balances out the charge on one magnesium ion: $Mg^{2+} + SO_4^{2-} = MgSO_4$.

Properties of Ionic Compounds

Ionic compounds such as sodium chloride and magnesium oxide have high melting and boiling points and conduct electricity when molten or in solution.

HT Sodium chloride consists of a giant lattice held together by the strong electrostatic forces of attraction between the positive (sodium) ions, and the negative (chloride) ions. These strong electrostatic forces are the ionic bonds.

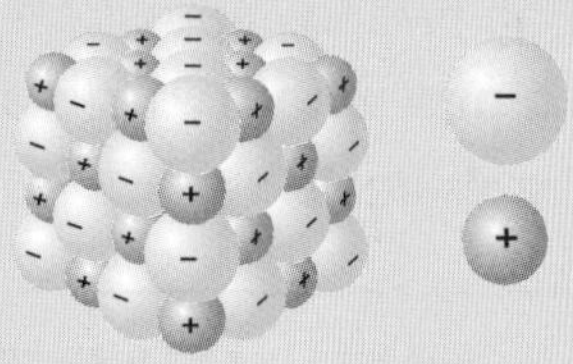
Negatively charged chloride ions
Positively charged sodium ions

Magnesium oxide consists of a similar giant lattice made up of Mg^{2+} and O^{2-} ions. Both of these compounds are examples of binary salts, which contain two elements, although all salts are made of positive and negative ions. Ionic compounds have high melting and boiling points due to the strong electrostatic forces of attraction that hold them together. They conduct electricity when molten or in solution because the charged ions are free to move between the electrodes and act as a current.

Solubility of Ionic Compounds

Some ionic compounds are soluble in water because the water can shield the electrostatic forces of attraction between the positive and negative ions. There are some general rules that can be used to help determine if an ionic compound is soluble in water.

The following are **soluble** in water.
- All carbonates of Group 1 compounds, e.g. K_2CO_3, Na_2CO_3.
- All nitrate compounds.
- Most common chloride compounds.
- Most common sulfate compounds.
- All alkalis, e.g. NaOH, $Ca(OH)_2$ (also known as slaked lime).

Solubility of Ionic Compounds (cont.)

The following are **insoluble** in water.

- Silver and lead halides, e.g. AgCl, $PbCl_2$.
- Sulfates of lead, barium and calcium, e.g. $BaSO_4$.
- Most common carbonate compounds, e.g. $CaCO_3$, $CuCO_3$.
- Most common hydroxide compounds, e.g. $Mg(OH)_2$ (also known as milk of magnesia).

Insoluble Salts

If the salt is formed by mixing two solutions is insoluble, it will form a solid **precipitate**. This is a **precipitation reaction**. For example:

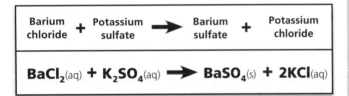

Barium chloride + Potassium sulfate → Barium sulfate + Potassium chloride

$$BaCl_2(aq) + K_2SO_4(aq) \rightarrow BaSO_4(s) + 2KCl(aq)$$

The pure solid salt can be separated by filtering the mixture, washing the solid residue in the filter paper with distilled water and drying the solid in a warm oven.

This can be demonstrated in the laboratory by mixing lead nitrate solution with sodium iodide solution.

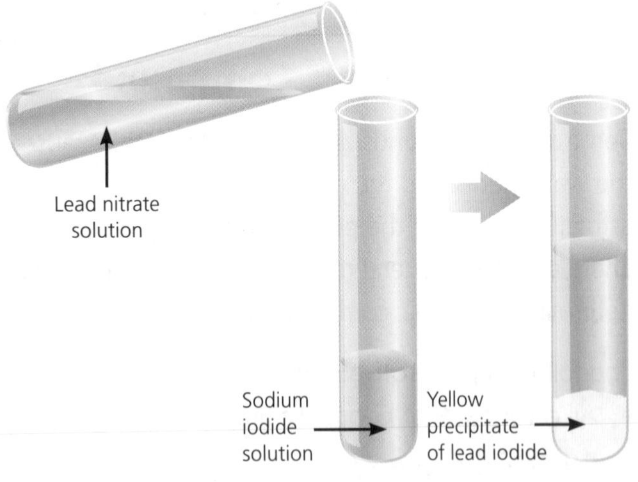
Lead nitrate solution
Sodium iodide solution
Yellow precipitate of lead iodide

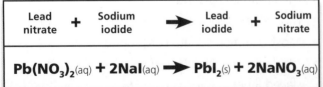

Lead nitrate + Sodium iodide → Lead iodide + Sodium nitrate

$$Pb(NO_3)_2(aq) + 2NaI(aq) \rightarrow PbI_2(s) + 2NaNO_3(aq)$$

The lead iodide mixture can then be filtered. The residue can then be washed with distilled water and dried.

Yellow lead iodide suspended in solution of sodium nitrate
Yellow lead iodide particles left behind on filter paper
Sodium nitrate solution
Dry yellow solid of lead iodide on filter paper

Remember that lead chloride and silver chloride are insoluble. The sulfates of barium, calcium and lead are also insoluble.

Barium Sulfate

Barium sulfate ($BaSO_4$) is extremely insoluble in water. This allows it to be used both in the laboratory and in medicine. In the laboratory it is the formation of barium sulfate that identifies sulfate ions in ionic substances.

In medicine, barium sulfate is used for barium meals, which are fed to X-ray patients. Soft tissue does not show up sufficiently well for diagnoses on plain X-rays. Barium sulfate is opaque to X-rays so helps soft tissues such as the digestive tract to become visible.

Barium salts in general are toxic if ingested. This is because they are soluble and can be absorbed by the body. In comparison, barium sulfate is considered non-toxic because it is highly insoluble and is not absorbed into the bloodstream.

Identifying Ions

An ionic substance can be identified by determining each type of **ion** within it. Each ion has its own unique test so this makes it easy to identify.

We can find out exactly what makes up a particular substance and subsequently determine whether or not the substance is harmful.

Test for Chloride Ions

To test for **chloride ions**, dilute nitric acid and silver nitrate solution are added to a solution of the ionic substance. If the ionic substance contains chloride ions then a white precipitate will be made.

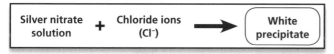

Silver nitrate solution	**+**	**Chloride ions (Cl^-)**	**→**	**White precipitate**

For example:

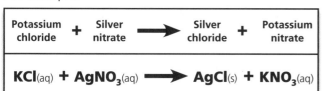

Potassium chloride	**+**	**Silver nitrate**	**→**	**Silver chloride**	**+**	**Potassium nitrate**

$$KCl(aq) + AgNO_3(aq) \rightarrow AgCl(s) + KNO_3(aq)$$

Tests for Compound Ions

Compound ions are ions that contain atoms of more than one element. These ions can be identified.

Identifying Carbonate Ions (CO_3^{2-})

Carbonate ions are compound ions. Carbonates (CO_3^{2-}) react with dilute acids to produce carbon dioxide. Therefore, if the unknown substance is a carbonate, it releases carbon dioxide when dilute acid is added to it. The CO_2 can be identified by testing with limewater.

For example:

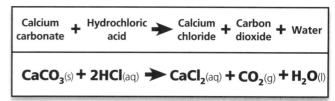

Calcium carbonate	**+**	**Hydrochloric acid**	**→**	**Calcium chloride**	**+**	**Carbon dioxide**	**+**	**Water**

$$CaCO_3(s) + 2HCl(aq) \rightarrow CaCl_2(aq) + CO_2(g) + H_2O(l)$$

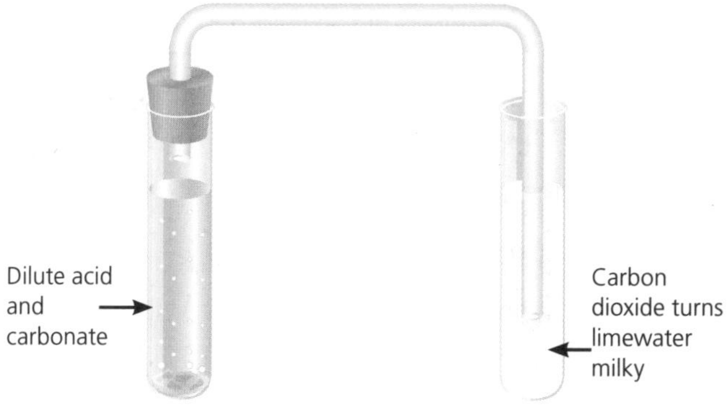
Dilute acid and carbonate
Carbon dioxide turns limewater milky

Identifying Sulfate Ions (SO_4^{2-})

When dilute hydrochloric acid and barium chloride solution are added to **SO_4^{2-} ions**, a white precipitate forms.

Therefore, if dilute hydrochloric acid and barium chloride solution are added to a solution of the unknown substance and a white precipitate forms, the substance contains SO_4^{2-} ions:

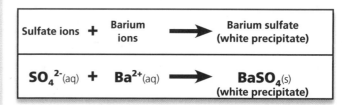

Sulfate ions	**+**	**Barium ions**	**→**	**Barium sulfate (white precipitate)**
$SO_4^{2-}(aq)$	**+**	$Ba^{2+}(aq)$	**→**	$BaSO_4(s)$ **(white precipitate)**

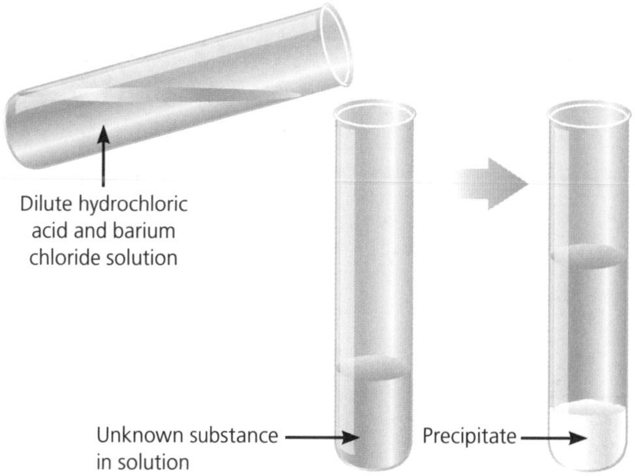
Dilute hydrochloric acid and barium chloride solution
Unknown substance in solution
Precipitate

Using Flames to Identify Ions

When held in a Bunsen flame, compounds of different metals produce different flame colours:

- sodium (Na^+) = yellow flame
- potassium (K^+) = lilac flame
- calcium (Ca^{2+}) = brick-red flame
- copper (Cu^{2+}) = blue–green flame.

Therefore, metal ions can easily be identified by heating a paste of the unknown ionic substance in a hot (blue) Bunsen flame.

To carry out a flame test to test for these metal ions, a paste of the solid compound is made with a small amount of hydrochloric acid. The end of the flame test wire (a nichrome wire) is dipped in the paste and held in the Bunsen flame (see also page 55). If the flame glows with a colour that is associated with a particular metal, the ions of that metal must be present in the substance.

When Bunsen and Kirchhoff viewed a sodium flame through a spectroscope, they observed bright lines in the yellow part of the visible spectrum.

Using spectroscopy to view other metals in their compounds this way showed that each metal had a distinct fingerprint in the spectrum and led to the discovery of new elements, including rubidium and caesium.

A 19th Century Spectroscope

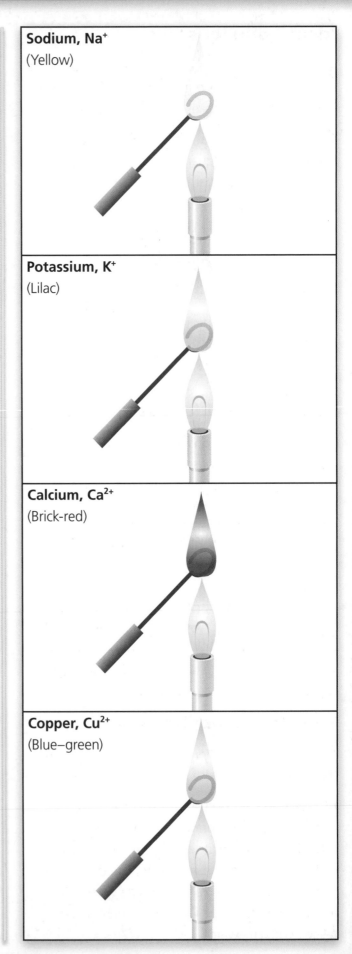

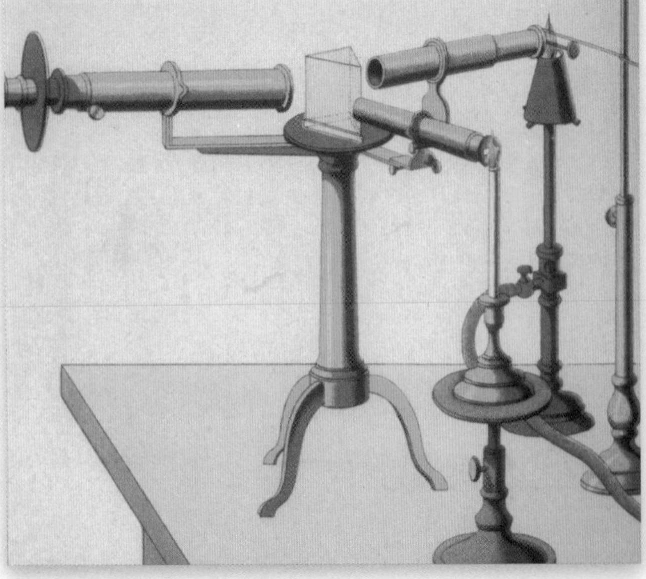

Sodium, Na^+
(Yellow)

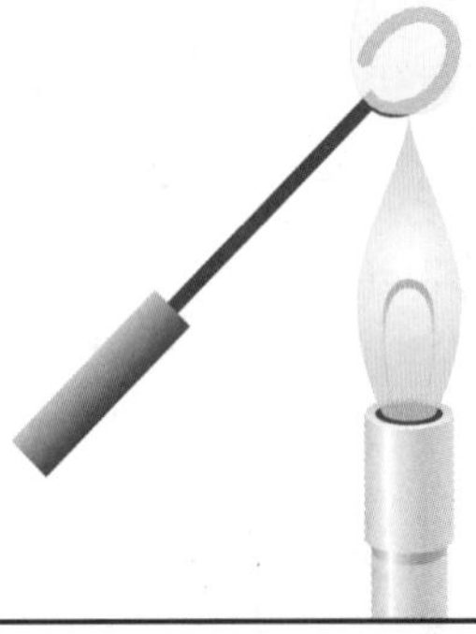

Potassium, K^+
(Lilac)

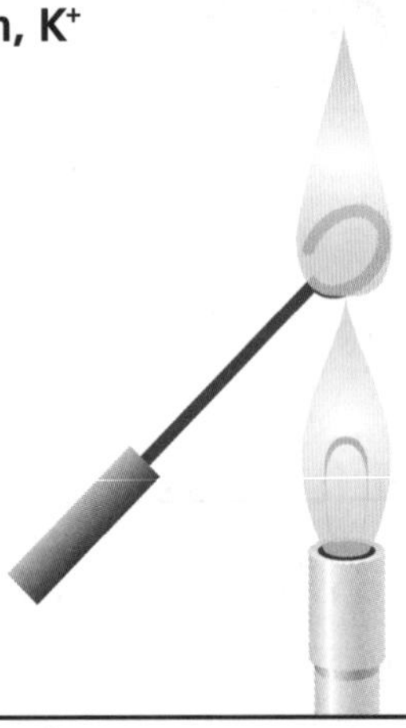

Calcium, Ca^{2+}
(Brick-red)

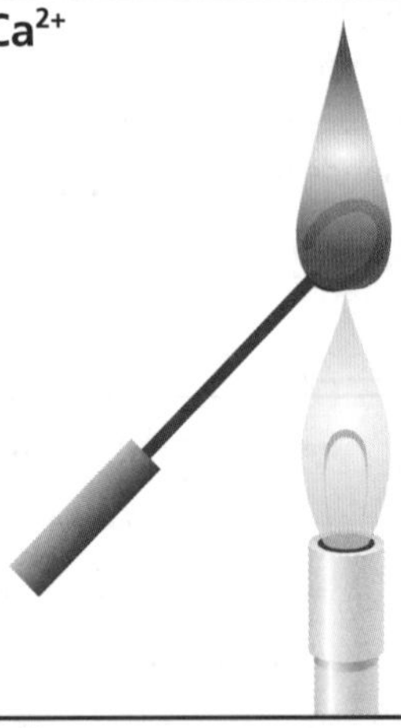

Copper, Cu^{2+}
(Blue–green)

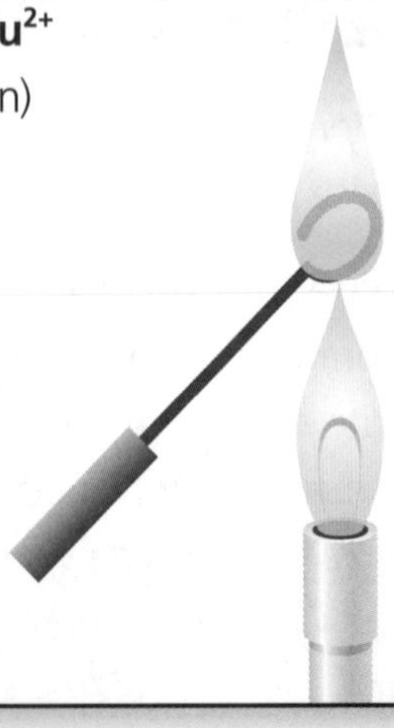

C2 Topic 3: Covalent Compounds and Separation Techniques

This topic looks at:
- how non-metals form molecules and compounds
- the properties of covalent compounds
- the properties and uses of diamond and graphite
- how immiscible and miscible liquids are separated
- what chromatography is and how it can be used

The Covalent Bond

A **covalent bond** occurs between non-metal atoms. The atoms share electrons in order to complete their outer shells. A covalent bond can occur between atoms of the same element or atoms of different elements. It results in the formation of **molecules**.

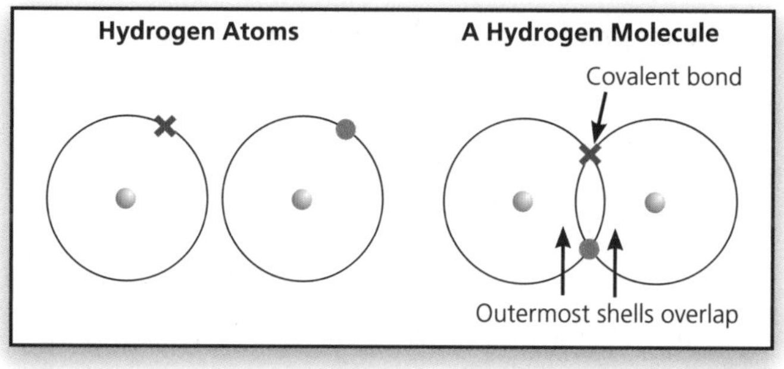
Hydrogen Atoms
A Hydrogen Molecule
Covalent bond
Outermost shells overlap

A **single covalent bond** is formed when two atoms share one pair of electrons. Each atom shares one electron in the bond, as in the example above – a hydrogen molecule.

If two pairs of electrons are shared, a **double bond** is formed. Each atom shares two of its electrons in the bond.

Examples

Hydrogen atoms and chlorine atoms can join together to form hydrogen chloride (HCl).

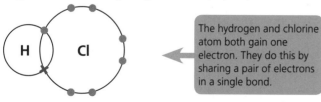

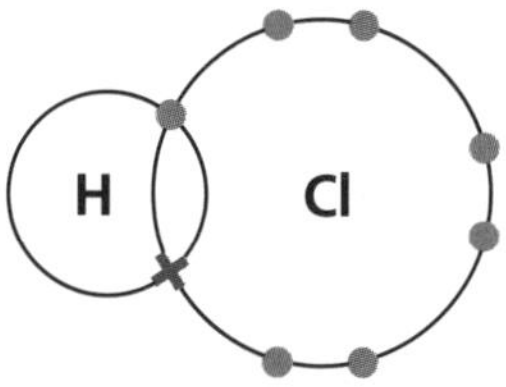
H
Cl

The hydrogen and chlorine atom both gain one electron. They do this by sharing a pair of electrons in a single bond.

Hydrogen atoms and oxygen atoms can join together to form water (H_2O).

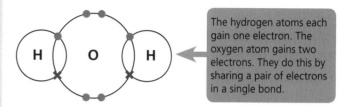

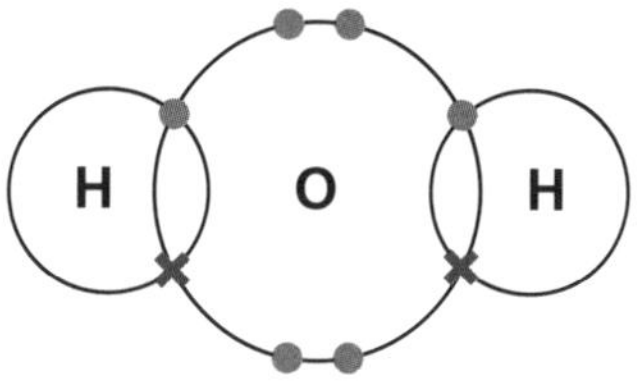
H
O
H

The hydrogen atoms each gain one electron. The oxygen atom gains two electrons. They do this by sharing a pair of electrons in a single bond.

Carbon atoms and hydrogen atoms can join together to form methane (CH_4).

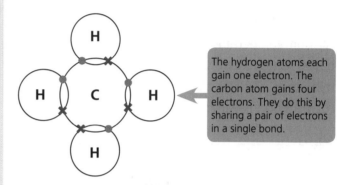

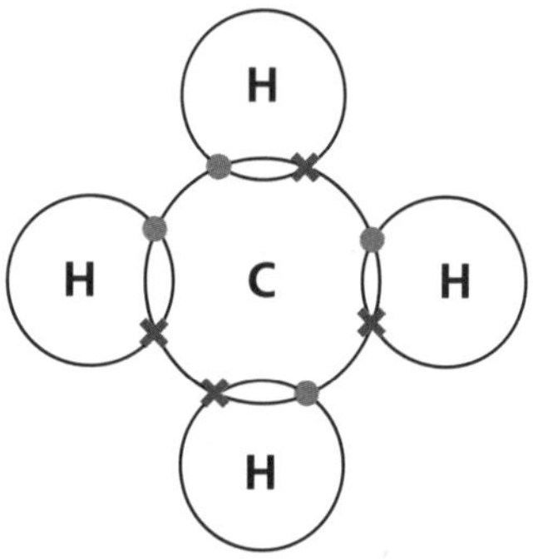
H
H
C
H
H

The hydrogen atoms each gain one electron. The carbon atom gains four electrons. They do this by sharing a pair of electrons in a single bond.

HT Both oxygen atoms achieve full outer shells by sharing electrons. They do this by sharing two pairs of electrons in a double covalent bond.

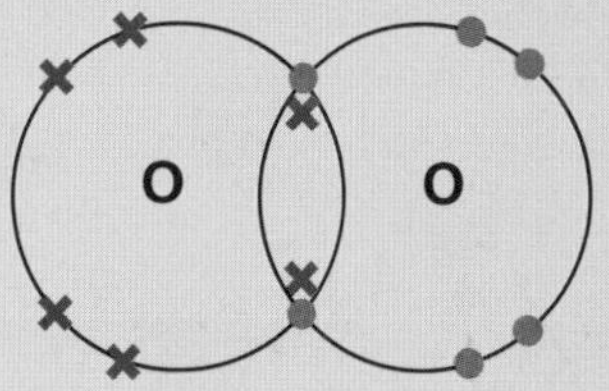
O
O

Carbon atoms and oxygen atoms can join together to form carbon dioxide (CO_2). The oxygen atoms each need two electrons to achieve a full outer shell. The carbon atom needs four electrons to achieve a full outer shell. They do this by sharing two pairs of electrons in a double covalent bond.

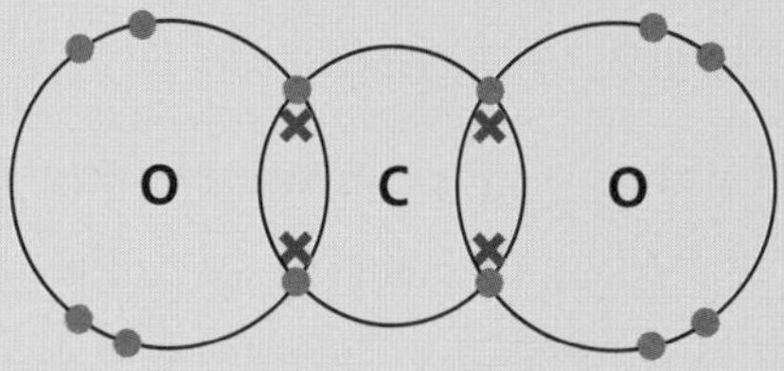
O
C
O

C2 Covalent Compounds and Separation Techniques

Properties of Simple Covalent Molecules

Substances with simple covalent structures consist of small molecules containing relatively few atoms. There are strong bonds between the atoms in the molecules, but there are weak forces between the molecules (weak inter-molecular forces).

This means that substances containing simple molecules have low melting and boiling points and have no overall charge so they cannot conduct electricity.

At room temperature many substances exist as gases, usually made up of molecules consisting of more than one atom. Again, there are strong covalent bonds within molecules but virtually no force of attraction between them. For example:

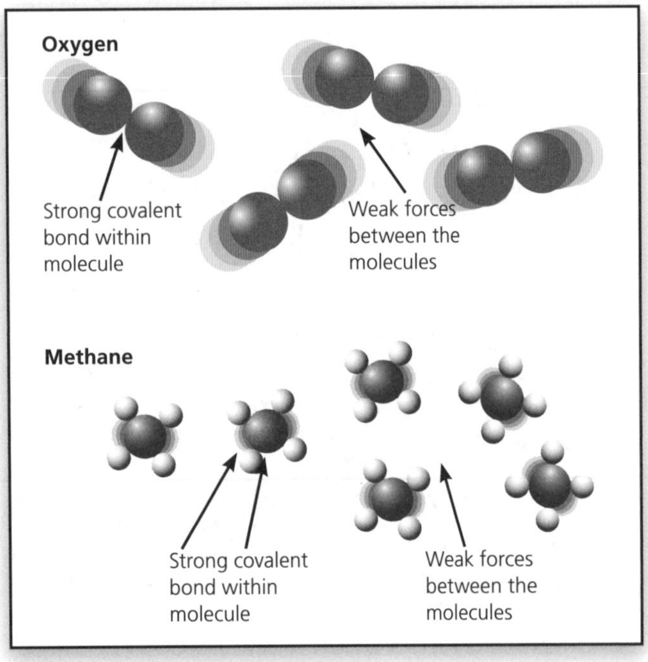

Oxygen
Strong covalent bond within molecule
Weak forces between the molecules
Methane
Strong covalent bond within molecule
Weak forces between the molecules

Covalent Structures – Giant Molecules

Giant covalent structures have many atoms joined to each other covalently throughout the whole structure. This makes their properties very different from those of simple covalently bonded molecules.

For example, each carbon atom in the structures of diamond and graphite shares its electrons with the atom next to it. This gives diamond and graphite giant regular structures with high melting and boiling points.

HT

Diamond (A Form of Carbon)

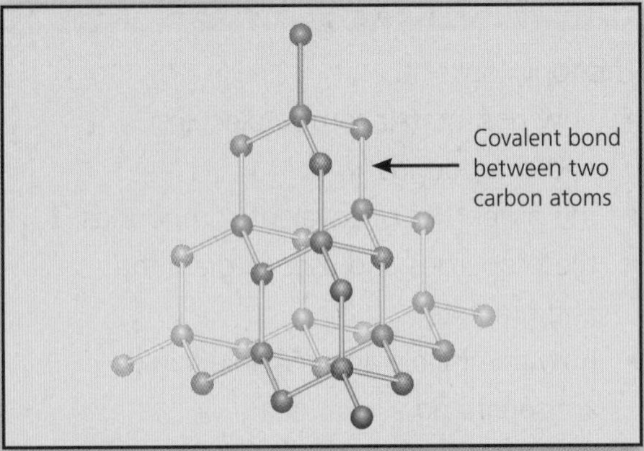

Covalent bond between two carbon atoms

Diamond is a giant, rigid, covalent structure (lattice) in which each carbon atom forms four covalent bonds with other carbon atoms. The strength of the covalent bonds results in diamond having very high melting and boiling points, which makes diamond very hard but unable to conduct electricity. This makes diamond a useful substance for making cutting tools.

Graphite (A Form of Carbon)

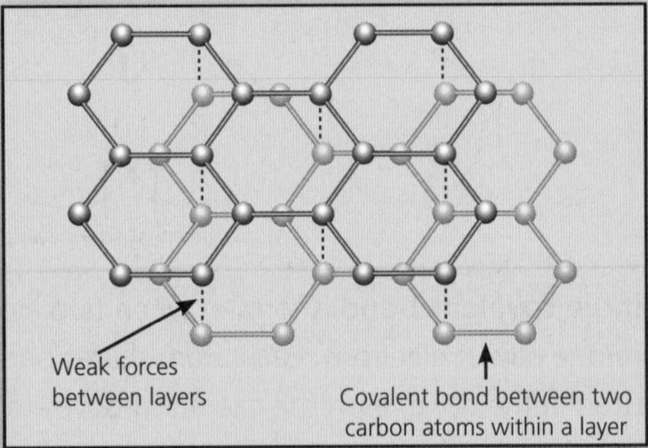

Weak forces between layers
Covalent bond between two carbon atoms within a layer

Graphite is a giant, covalent, layered structure. The layers can slide past each other, making graphite soft and slippery. Like diamond, graphite has high melting and boiling points. There are weak forces of attraction between layers because each carbon atom forms only three covalent bonds within the layers, so one electron from each carbon atom can be delocalised (moved). This allows graphite to conduct heat and electricity. Graphite is used to make electrodes and lubricants.

Testing Different Compounds

It is possible to demonstrate the differences in properties between some ionic and some covalent compounds by:

- trying to melt them, using a Bunsen burner
- placing electrodes into a sample of the substance to see if it conducts electricity as a solid, liquid or aqueous solution
- timing how long it takes the substance to dissolve in water.

You may have done these tests with sodium chloride, magnesium sulfate, sucrose (sugar) and some other compounds.

Separating Mixtures

Solids can form mixtures and if one of the solids is soluble in water then the mixture can easily be separated by filtration. Liquids can also be mixed together. They too can be separated.

When two liquids, such as oil and water, are mixed together and form distinct separate layers they are called **immiscible** (they cannot mix together). In this example, the water can be separated from the oil using a separating funnel as shown in the diagram opposite.

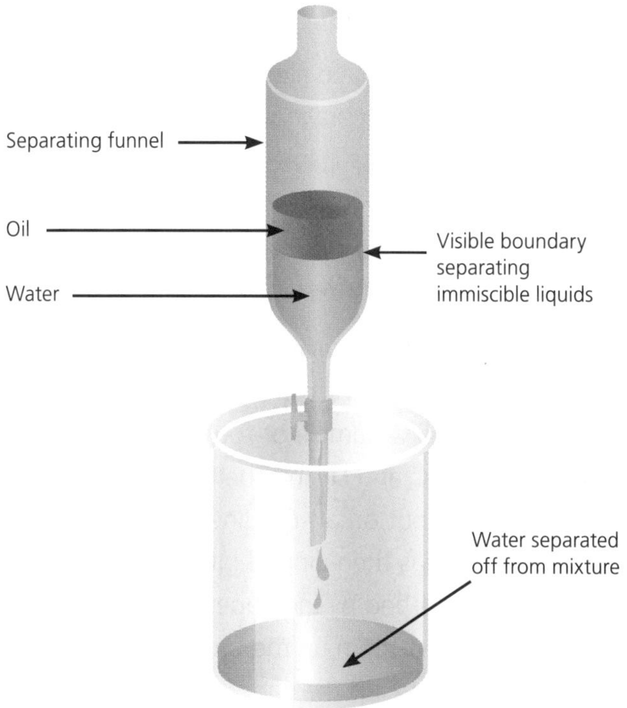
Separating funnel
Oil
Water
Visible boundary separating immiscible liquids
Water separated off from mixture

When liquids are mixed together and they stay mixed together they are called **miscible**. In this case the liquids have to be separated by a different method. They are separated using distillation or fractional distillation. For example, liquid air is a mixture of gases that have been cooled to a liquid and then separated by fractional distillation.

Fractional Distillation of Air

Dry air is made up of gases, including oxygen, nitrogen and carbon dioxide. Both oxygen and nitrogen can be obtained by separating them from liquid air by means of **fractional distillation**. We use oxygen in steel-making and nitrogen for freezing food quickly.

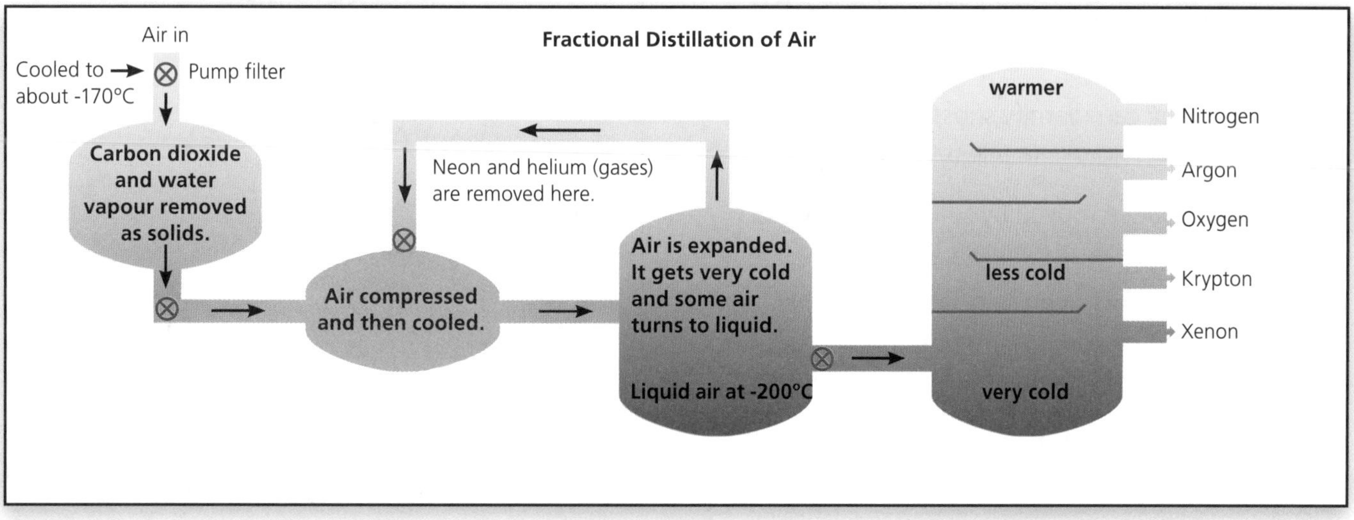
Fractional Distillation of Air
Air in
Cooled to about -170°C
Pump filter
Carbon dioxide and water vapour removed as solids.
Air compressed and then cooled.
Neon and helium (gases) are removed here.
Air is expanded. It gets very cold and some air turns to liquid.
Liquid air at -200°C
warmer
less cold
very cold
Nitrogen
Argon
Oxygen
Krypton
Xenon

Chromatography

Evaporation, distillation and filtration are not always suitable methods for separating mixtures. For example, they are all unsuitable for separating the small amount of dissolved dyes used in food colourings. To separate such mixtures, a process called **chromatography** would have to be used.

Chromatography allows unknown components in a mixture to be identified by comparing them to known substances. A sample of four known substances (A, B, C and D) and the unknown substance (X) are put on a 'start line' on a piece of filter paper, which is then dipped into a solvent. As the solvent is absorbed by the paper it dissolves the samples and carries them up the paper. The substances will move up the paper at different rates because the different substances have different levels of solubility. The unknown substance (X) can be identified by comparing the horizontal spots.

The more soluble the substance is, then the further up the paper the substance is carried by the solvent. For paper chromatography this is usually water. Other solvents used in chromatography include ethanol.

Different methods of chromatography have been developed, resulting in a wide range of uses in the medical, chemical, food and forensics professions. Chromatography can be used to check foodstuffs for additives, contaminants and other substances. It can also be used by forensic investigators to separate traces of chemicals in house or building fires and to identify seized drugs.

Chromatography can identify substances by working out their R_f values using the following formula:

$$R_f \text{ value} = \frac{\text{distance moved by soluble substance}}{\text{distance moved by solvent}}$$

Once the R_f value has been calculated, the substance can be identified by comparison to published tables. An unknown amino acid in a food can easily be identified this way, or an unknown chemical found at the scene of a house fire.

Apparatus For Chromatography

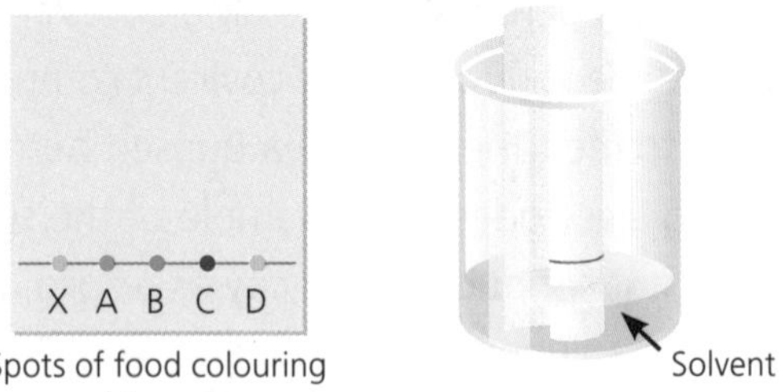
X A B C D
Solvent

Spots of food colouring on filter paper

By comparing food colourings A, B, C and D to substance X, we can see that substance X is food colouring D.

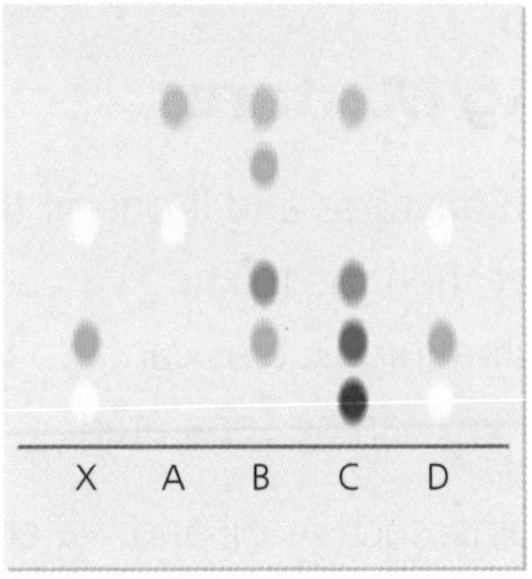
X A B C D

A chromatogram showing how the food colourings have split into their dyes

C2 Topic 4: Groups in the Periodic Table

This topic looks at:

- the structure and properties of metals
- the transition metals and their typical properties
- the alkali metals and their typical properties and reactions
- the halogens and their typical properties and reactions
- the noble gases, how they were discovered and their properties and uses
- how elements and compounds are classified and how their properties differ according to their structure

Metals

Metal atoms form giant crystalline structures. The atoms are packed tightly together so the outer electrons become separated from the atom. The result is a lattice structure of positive ions in a sea of free-moving electrons.

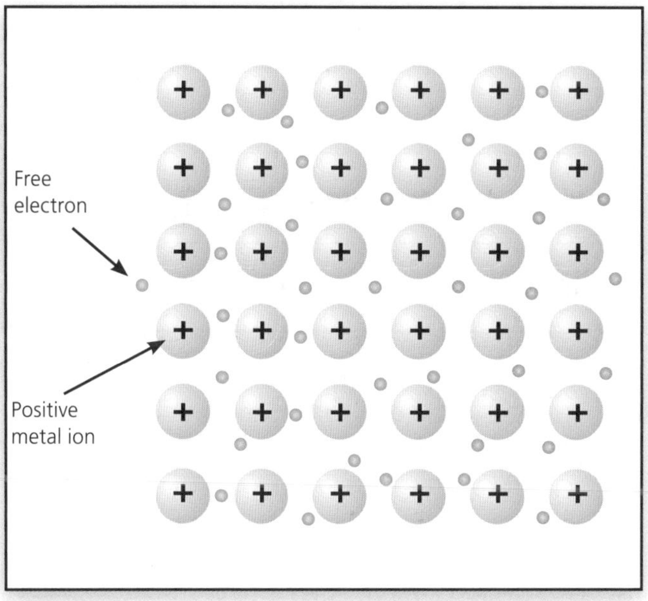
Free electron
Positive metal ion

Metals are very good conductors of electricity because their outer shell electrons can move freely within the structure, carrying the electric charge.

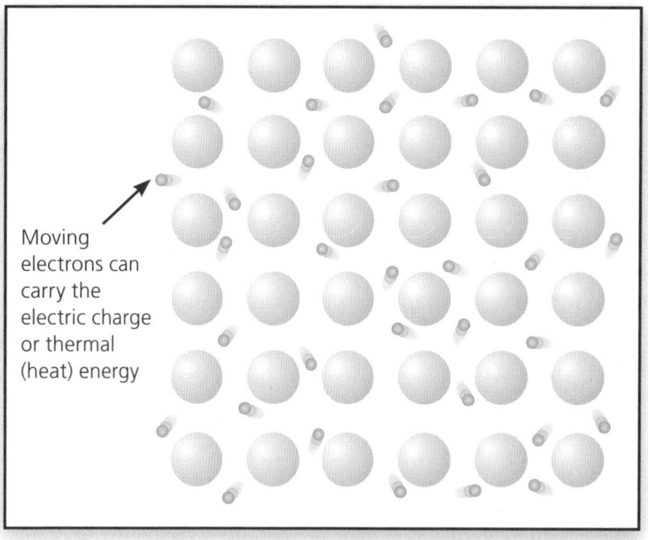
Moving electrons can carry the electric charge or thermal (heat) energy

Metal ions align in their structure in layers, which gives rise to their malleability. The layers can easily slide over each other.

The Transition Metals

The **transition metals** are included in the central block between Group 2 and Group 3 of the periodic table.

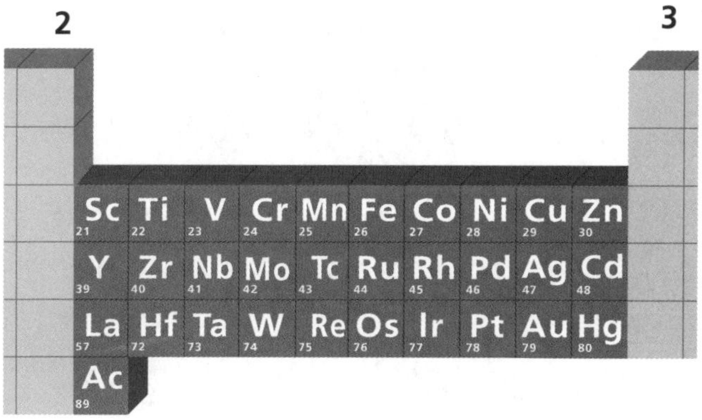
2
3
Sc 21 Ti 22 V 23 Cr 24 Mn 25 Fe 26 Co 27 Ni 28 Cu 29 Zn 30
Y 39 Zr 40 Nb 41 Mo 42 Tc 43 Ru 44 Rh 45 Pd 46 Ag 47 Cd 48
La 57 Hf 72 Ta 73 W 74 Re 75 Os 76 Ir 77 Pt 78 Au 79 Hg 80
Ac 89

Iron and copper are well-known examples of transition elements.

Copper – A Transition Metal – Used to Make Pipes

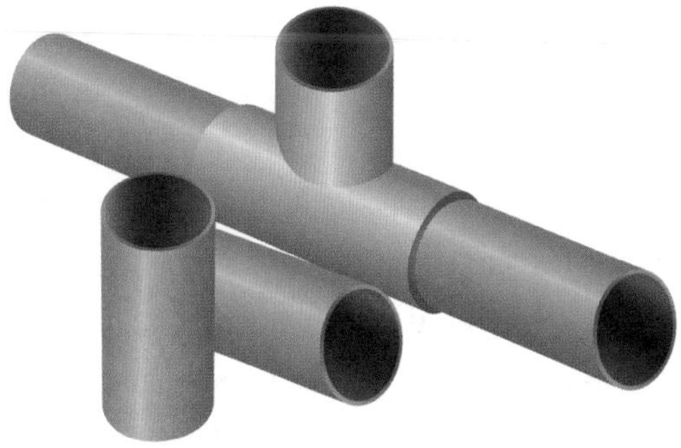

C2 Groups in the Periodic Table

Properties of the Transition Metals

- Transition metals are malleable, ductile and dense and have high melting points.
- They are good conductors of heat and electricity.
- Transition metal compounds are often coloured; for example, copper sulfate crystals are blue.

The Alkali Metals

The **alkali metals** occupy the first vertical group (Group 1) at the left-hand side of the periodic table. Lithium, sodium and potassium are typical members of this group.

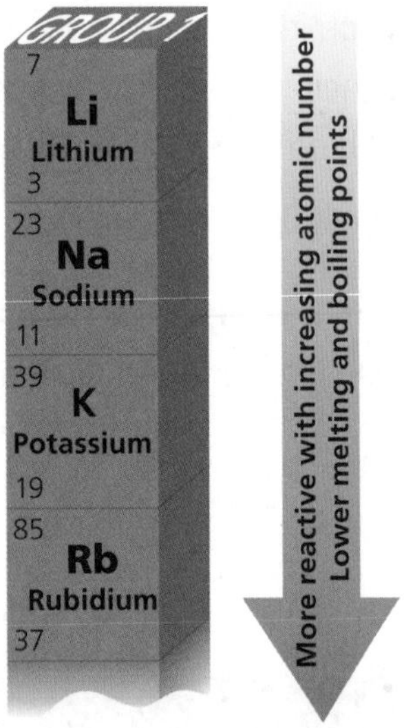
GROUP 1
7
Li
Lithium
3
23
Na
Sodium
11
39
K
Potassium
19
85
Rb
Rubidium
37
More reactive with increasing atomic number
Lower melting and boiling points

Properties of the Alkali Metals

- They are soft and have low melting points (e.g. potassium has a melting point of 63°C).
- Their reactions with water become increasingly exothermic (increase in temperature).
- Their high degree of reactivity means they must be stored under oil.
- The further down Group 1 the metal is, the greater its reactivity.
- The further down Group 1 the metal is, the further away the lone outer electron is from the nucleus.
- Francium is the most reactive alkali metal: its nucleus is unstable and it is a radioactive element.
- They react vigorously with water to form hydroxides, which are alkaline (higher than 7 on the pH scale), and hydrogen gas.

A simple test can be performed for hydrogen gas. Hydrogen collected in the inverted test tube makes a squeaky pop when lit.

Pop!
Test tube
of hydrogen
Lighted
splint

Reactions of the Alkali Metals

On contact with water, lithium floats, begins to move around and fizzes.

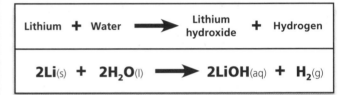

Lithium + Water → Lithium hydroxide + Hydrogen
$2Li_{(s)} + 2H_2O_{(l)} \longrightarrow 2LiOH_{(aq)} + H_{2(g)}$

On contact with water, sodium shoots across the water surface and fizzes vigorously.

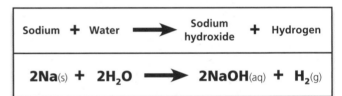

Sodium + Water → Sodium hydroxide + Hydrogen
$2Na_{(s)} + 2H_2O \longrightarrow 2NaOH_{(aq)} + H_{2(g)}$

Potassium has a more violent reaction as the gas ignites whilst the metal moves very quickly across the surface of the water.

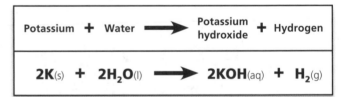

Potassium + Water → Potassium hydroxide + Hydrogen
$2K_{(s)} + 2H_2O_{(l)} \longrightarrow 2KOH_{(aq)} + H_{2(g)}$

HT Reactivity of Group 1 Alkali Metals

Alkali metals react to lose their outer electron and gain a full outer shell. The smaller the atom, the greater the effect of the attractive force of the protons in the nucleus, so the lone outer electron is held in place. As the atoms get bigger this effect decreases. The outer electron is lost more easily and the metal is therefore more reactive.

Halogens

The **halogens** are found in Group 7 of the periodic table. There are five non-metals in Group 7; the top four are the ones you need to remember. They are all different in colour. Their melting and boiling points determine their physical state at room temperature. The halogens all exist as diatomic molecules.

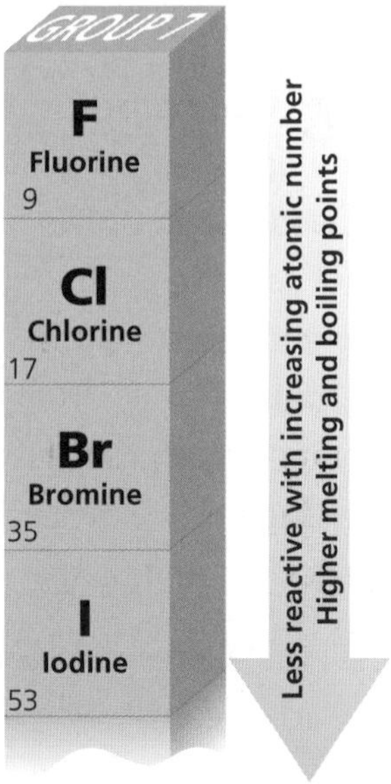
GROUP 7
F
Fluorine
9
Cl
Chlorine
17
Br
Bromine
35
I
Iodine
53
Less reactive with increasing atomic number
Higher melting and boiling points

Halogen	Fluorine	Chlorine	Bromine	Iodine
Boiling Point (°C)	-188°C	-34°C	59°C (melting point -7°C)	187°C (melting point 114°C)
Colour and Physical State at Room Temperature	Pale yellow gas	Pale green gas	Red-brown liquid	Dark grey solid

Displacement Reactions

The halogens' atomic numbers increase as we go down the group and they become less reactive. This can be shown by the **displacement reactions** of halogens with solutions of other **halides**. In summary:

- fluorine is the most reactive, followed by chlorine
- bromine and iodine are the least reactive.

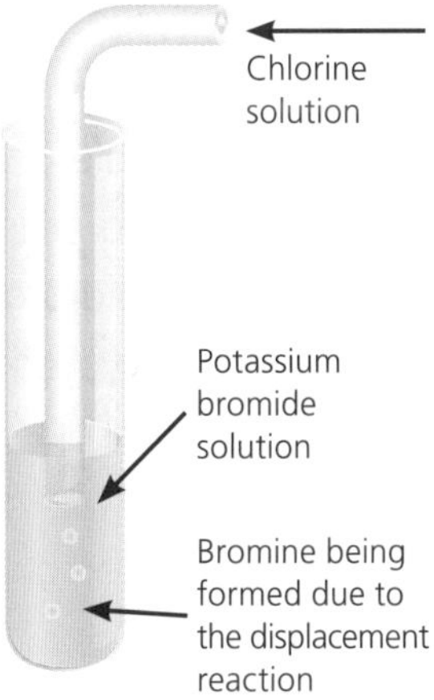
Chlorine solution
Potassium bromide solution
Bromine being formed due to the displacement reaction

	Potassium Chloride	Potassium Bromide	Potassium Iodide
Fluorine F_2	Potassium fluoride and chlorine	Potassium fluoride and bromine	Potassium fluoride and iodine
Chlorine Cl_2	✕	Potassium chloride and bromine	Potassium chloride and iodine
Bromine Br_2	No reaction	✕	Potassium bromide and iodine
Iodine I_2	No reaction	No reaction	✕

If there is a displacement reaction then it will follow this pattern.

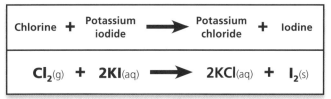

Chlorine + Potassium iodide → Potassium chloride + Iodine

$$Cl_2(g) + 2KI(aq) \longrightarrow 2KCl(aq) + I_2(s)$$

Uses of the Halogens

Chlorine is used to kill bacteria, for example, in swimming pools and domestic water supplies. It is also used for bleaching paper, wood and cloth. Iodine solution is used as an antiseptic.

Reactions of Halogens and Iron

Iron, in the form of iron wool, is heated strongly and **chlorine** gas is passed over it in a fume cupboard. The iron wool will glow brightly as the following reaction takes place.

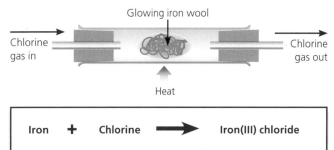

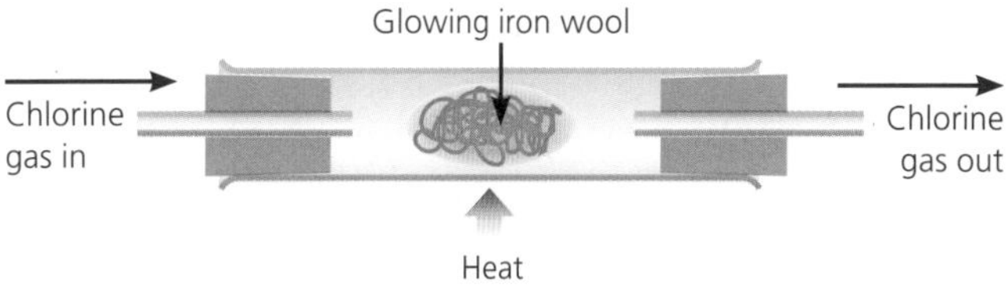
Glowing iron wool
Chlorine gas in
Chlorine gas out
Heat

Iron + Chlorine → Iron(III) chloride

$$2Fe(s) + 3Cl_2(g) \longrightarrow 2FeCl_3(s)$$

Chlorine will react with other metals to form chlorides. For example:

Sodium + Chlorine → Sodium chloride

$$2Na(s) + Cl_2(g) \longrightarrow 2NaCl(s)$$

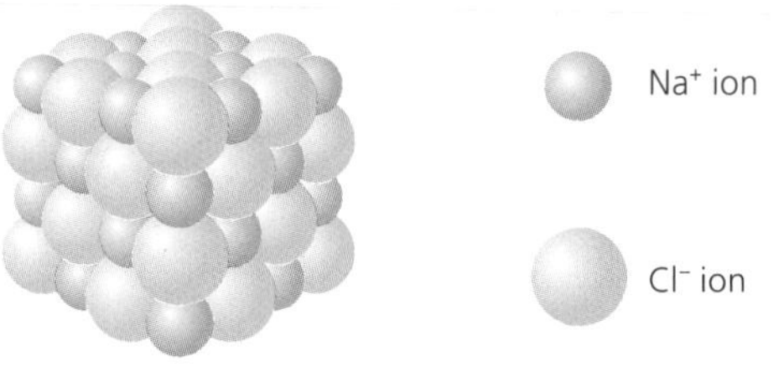
Na⁺ ion
Cl⁻ ion

So, one atom of sodium plus one atom of chlorine produces sodium chloride. Bromine will form bromides and iodine will form iodides.

Reactions of Halogens with Hydrogen

The halogens will react with hydrogen to give hydrogen halides.

Hydrogen halides will form colourless acidic gases which are highly soluble in water. The resultant solution is acidic. For example:

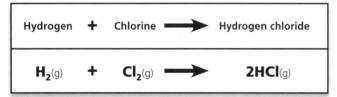

Hydrogen	**+**	**Chlorine**	**→**	**Hydrogen chloride**
$H_2(g)$	**+**	$Cl_2(g)$	**→**	$2HCl(g)$

Then dissolve this gas in water.

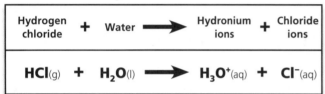

Hydrogen chloride	**+**	**Water**	**→**	**Hydronium ions**	**+**	**Chloride ions**
$HCl(g)$	**+**	$H_2O(l)$	**→**	$H_3O^+(aq)$	**+**	$Cl^-(aq)$

Discovery of Noble Gases

In the late 1880s Sir William Ramsey recorded a visible spectrum for an unknown gas that matched and confirmed the earlier discovery of **helium**.

Argon was discovered in 1895 when Rayleigh and Ramsey measured the densities of gases and compared nitrogen obtained from ammonia to nitrogen left when other gases were removed from air. A visible spectrum told them that this was a previously unknown gas, which they named argon.

Neon was discovered in 1898 when Ramsey and Morris Travers discovered other lines in the visible spectrum for argon.

Again, visible spectra were used to identify **krypton**, when known constituents were boiled off from a sample of air.

Xenon was found when it glowed bright blue in a discharge tube.

Scientists and chemists were able to use patterns in the physical properties of the noble gases, such as boiling point or density, to estimate values for other members of the group.

The Noble Gases

The **noble gases** are located in a vertical group at the right-hand side of the periodic table. This is called Group 0. The noble gases have no smell and are colourless. They glow with a particular colour when electricity passes through them. The noble gases generally do not react due to their full outer electron shell (a complete electronic arrangement). (Note that helium has two electrons in its outer shell.)

All the elements in Group 0 are **chemically inert**, which means they are unreactive.

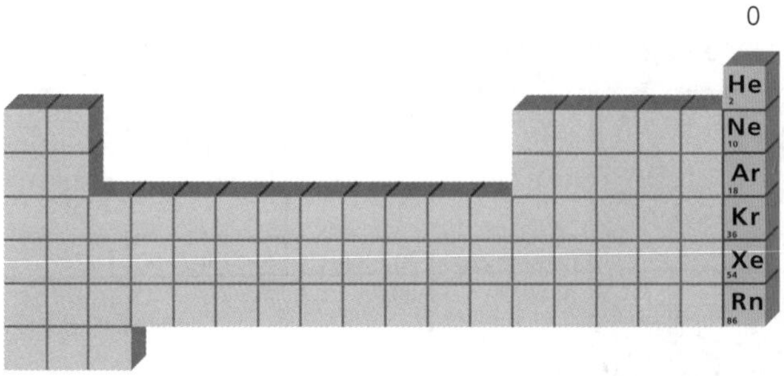
0
He
Ne
Ar
Kr
Xe
Rn

Uses of the Noble Gases

- Helium is used in airships and weather balloons because it is much less dense than air and is non-flammable.
- Argon is used in light bulbs because it is unreactive and provides an inert atmosphere.
- Argon is also used in welding, where it acts like a shield around the weld until the metal cools, preventing any reaction with oxygen in the air.
- Argon, krypton and neon are all used in fluorescent lights and discharge tubes.

Summary of the Different Types of Structure and Bonding

Ionic Compounds

There are millions of different substances in existence. Some compounds are made up of ions. These substances are all solids at room temperature. The ions can be positively charged (metals and hydrogen) or negatively charged (non-metals). The oppositely charged ions will be attracted by a strong electrostatic attraction, which is known as an ionic bond. It is the strong attractive forces between the oppositely charged ions that give the ionic compounds many of their properties.

Molecular Compounds

Some substances are made up of molecules. These molecules are said to have a simple or small molecule structure and contain atoms that are joined together by covalent bonds. This type of bond involves the sharing of pairs of electrons between non-metal atoms.

The covalent bonds between the atoms within the molecules are very strong. There are no bonds between the molecules but the molecules are held in their structure by weak attractive forces between them, called intermolecular forces.

It is these weak intermolecular forces between the molecules that give rise to their low boiling and melting points and whether they are a solid, liquid or gas at room temperature.

Giant Covalent Structures

As well as forming small molecules, non-metal atoms can also form giant covalent structures in which all the atoms are joined together by covalent bonds in a massive network. The covalent bonds that keep these atoms together within their structure are (like these bonds in small molecules) very strong and give the substances their properties.

Metallic Structures

About three-quarters of the elements in the periodic table are metals. Metal atoms form a giant, regular structure held together by the strong electrostatic attractions of positive metal ions and the free-moving outer shell electrons. This metallic structure gives metals their chemical properties.

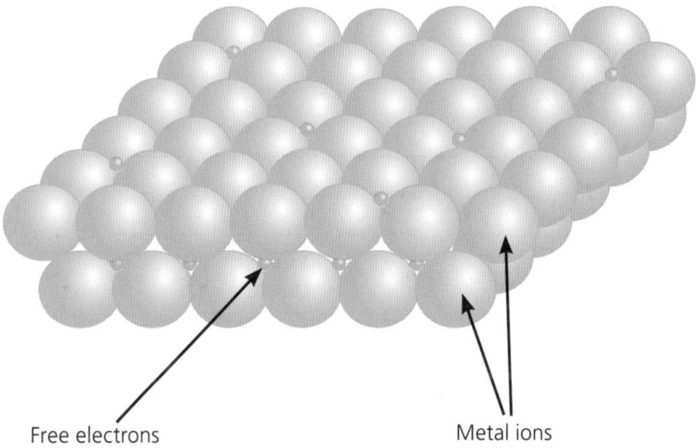
Free electrons
(negative charge)
Metal ions
(positive charge)

Each of these types of bonding and their resultant properties can be compared in the table below.

Bonding	Ionic (between metals and non-metals)	Covalent (between non-metals)		Metallic (between metals)
Structure	Giant ionic	Giant covalent	Simple molecular	Giant metallic
Melting and Boiling Point	High	High	Low	High
Soluble in Water	Yes	No	Yes	No
Conduct Electricity	As a solid, no; when molten or in solution, yes (ions are free to move)	No	No	Yes (free electrons)

C2 Topic 5: Chemical Reactions

This topic looks at:

- the difference between endothermic and exothermic reactions
- how reactions can differ in rate
- how collisions between particles affect a rate of reaction
- why collisions in chemical reactions are important
- how catalysts affect chemical reactions
- how a car's catalytic convertor works

Energy Changes in Chemical Reactions

A **chemical reaction** involves two or more **reactants** reacting together to make new substances, known as **products**. Reactants and products may be elements or compounds.

Reactants → Products
Sodium + Chlorine → Sodium chloride (+ Heat energy)
$2Na_{(s)} + Cl_{2(g)} \rightarrow 2NaCl_{(s)}$ (+ Heat energy)

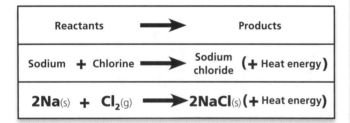

Temperature Changes in Reactions

All chemical reactions are accompanied by a temperature change.

Exothermic reactions are accompanied by a rise in temperature. Thermal (heat) energy is transferred out to the surroundings. Combustion is an example of an exothermic reaction.

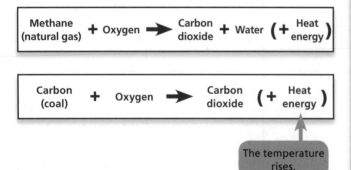

Methane (natural gas) + Oxygen → Carbon dioxide + Water (+ Heat energy)

Carbon (coal) + Oxygen → Carbon dioxide (+ Heat energy)

The temperature rises.

Other exothermic reactions include:

- respiration in body cells
- hydration of ethene to form ethanol
- neutralising alkalis with acids.

Endothermic reactions and changes are accompanied by a fall in temperature. Thermal energy is transferred in from the surroundings. Dissolving ammonium nitrate crystals in water is an example of an endothermic change.

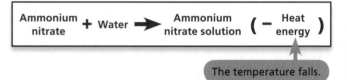

Ammonium nitrate + Water → Ammonium nitrate solution (− Heat energy)

The temperature falls.

Examples of endothermic reactions include:

- the reaction between citric acid and sodium hydrogencarbonate solution
- polymerisation of ethene to polyethene (see page 23)
- reduction of silver ions to silver in photography.

By carrying out some chemical reactions and measuring the temperature before and after, it is possible to see that chemical reactions can be endothermic or exothermic. For example:

- dissolving ethanoic acid in sodium carbonate is accompanied by a temperature drop
- reacting sodium hydroxide with hydrochloric acid is accompanied by a temperature rise
- displacing copper from a solution of copper sulfate, using zinc powder, is accompanied by a temperature rise.

Making and Breaking Bonds

Energy must be supplied to break chemical bonds.

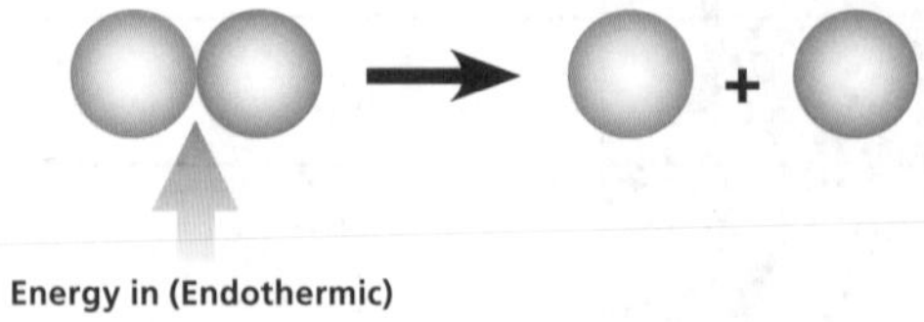
Energy in (Endothermic)

Energy is released when chemical bonds are made.

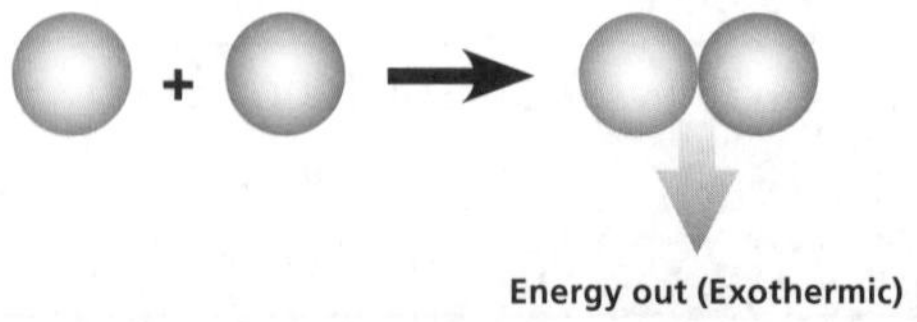
Energy out (Exothermic)

In order to produce new substances in a chemical reaction, the bonds in the reactants must be broken and new bonds must be made in the products.

If more energy is needed to break old bonds than is released when new bonds are made, the reaction is endothermic overall.

(Energy +) Reactants
Products
(− Energy)

If more energy is released when new bonds are made than is needed to break the old bonds, the reaction is exothermic overall.

(Energy +) Reactants
Products
(+ Energy)

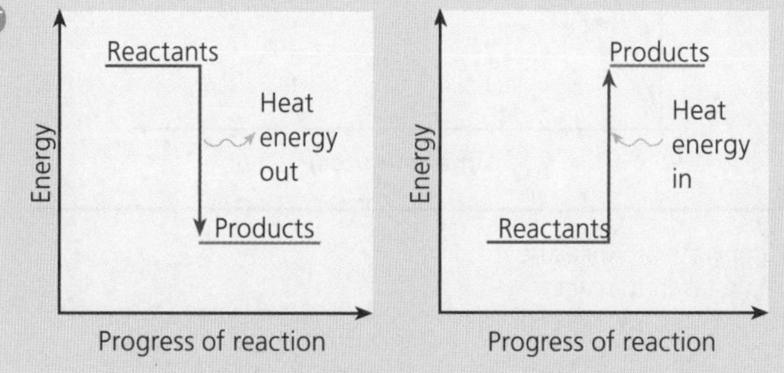
Reactants
Heat energy out
Products
Energy
Progress of reaction
Products
Heat energy in
Reactants
Energy
Progress of reaction

Exothermic reaction **Endothermic reaction**

Rates of Reaction

Chemical reactions happen at different rates (for example, within seconds, days or even years).

Rates of chemical reactions can be increased so that the reactions take place more quickly. This can be achieved by:

- increasing the **temperature** of the reactants so the particles collide more frequently
- increasing the **surface area** of the solid reactants so the particles collide more frequently
- increasing the **concentration** of one of the reactants so the particles collide more frequently.

Collision Theory

Chemical reactions usually occur when reacting particles collide with each other with sufficient energy to react.

All substances are made up of particles. These particles may be atoms, molecules or ions. In a chemical reaction these reactant particles collide with each other. Not all chemical reactions work because the particles, on colliding, may rebound and remain unchanged if they do not posses sufficient energy to break bonds.

Low Temperature

In a cold reaction mixture the particles are moving quite slowly – the particles collide with each other less often, with less energy, and fewer collisions are successful in a given time (per second).

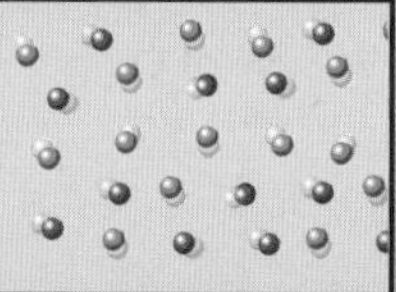

High Temperature

If we heat the reaction mixture, the particles move more quickly – the particles collide with each other more often, with greater energy, and many more collisions are successful in a given time (per second).

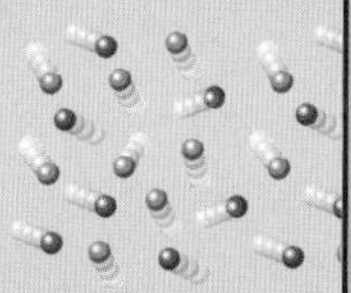

Small Surface Area

Large particles have a small surface area in relation to their volume – fewer particles are exposed and available for collisions. This results in fewer collisions per second and a slower reaction.

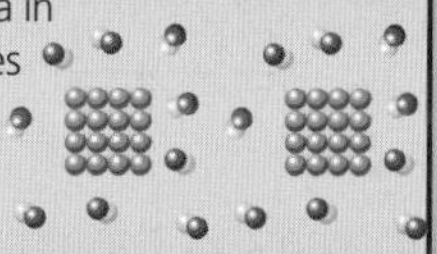

Large Surface Area

Small particles have a large surface area in relation to their volume – more particles are exposed and available for collisions. This results in more collisions per second and a faster reaction.

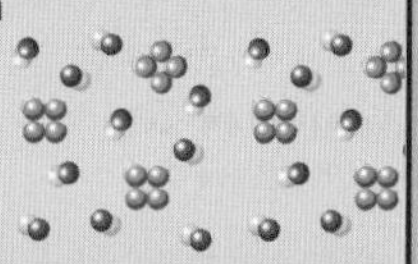

Low Concentration

In a reaction where one or both reactants are in low concentrations, the particles are spread out and collide with each other less often, resulting in fewer successful collisions per second.

High Concentration

In a reaction where one or both reactants are in high concentrations, the particles are packed closely together and collide with each other more often, resulting in more successful collisions per second.

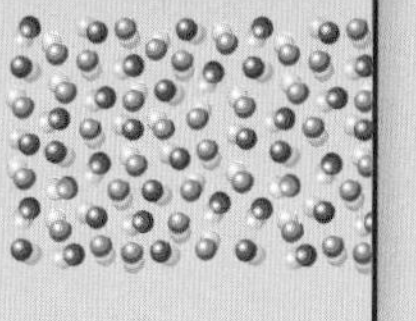

Examples of Changing Rates of Reactions

Calcium carbonate reacts with hydrochloric acid to produce calcium chloride, water and carbon dioxide.

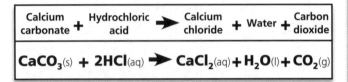

Calcium carbonate	**+**	**Hydrochloric acid**	**→**	**Calcium chloride**	**+**	**Water**	**+**	**Carbon dioxide**
$CaCO_3(s)$	+	$2HCl(aq)$	→	$CaCl_2(aq)$	+	$H_2O(l)$	+	$CO_2(g)$

We can measure how long it takes a given mass of calcium carbonate to react completely and see how changing factors can affect the rate of the reaction.

1. Heating the solution makes the reaction occur more quickly.

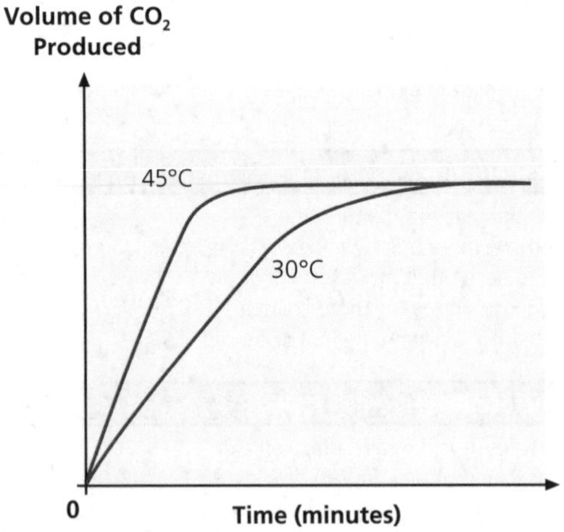
Volume of CO2 Produced
45°C
30°C
0
Time (minutes)

2. Increasing the concentration of the acid makes the reaction occur more quickly.

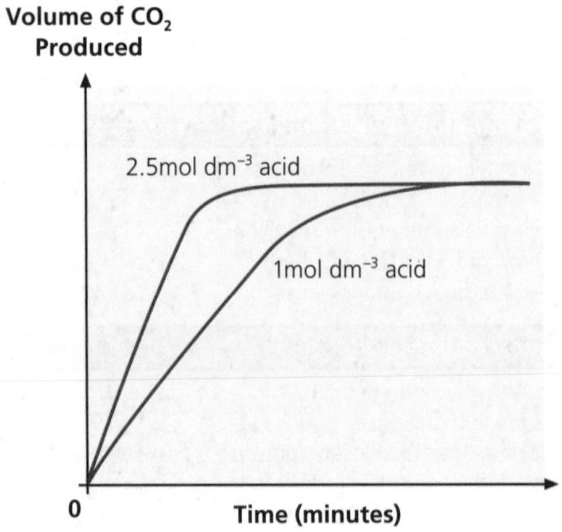
Volume of CO2 Produced
2.5mol dm-3 acid
1mol dm-3 acid
0
Time (minutes)

3. Making the solid have a large surface area increases the rate of the reaction. This can be seen by measuring the amount of carbon dioxide given off every minute. Calcium carbonate chips and then the same mass of finely crushed calcium carbonate can be used.

N.B. There is the same mass of calcium carbonate in both reactions, so the same volume of carbon dioxide is produced.

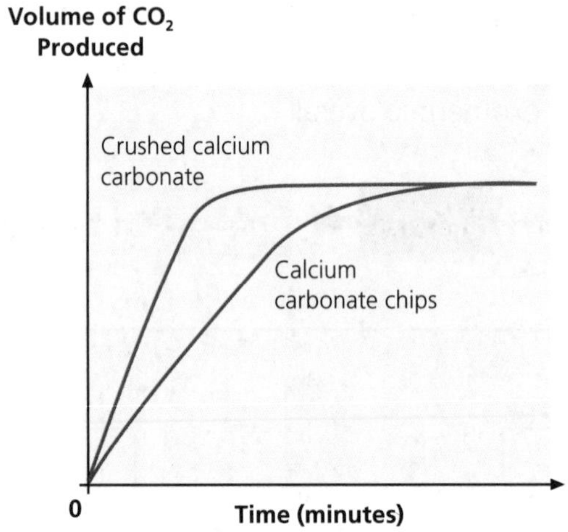
Volume of CO2 Produced
Crushed calcium carbonate
Calcium carbonate chips
0
Time (minutes)

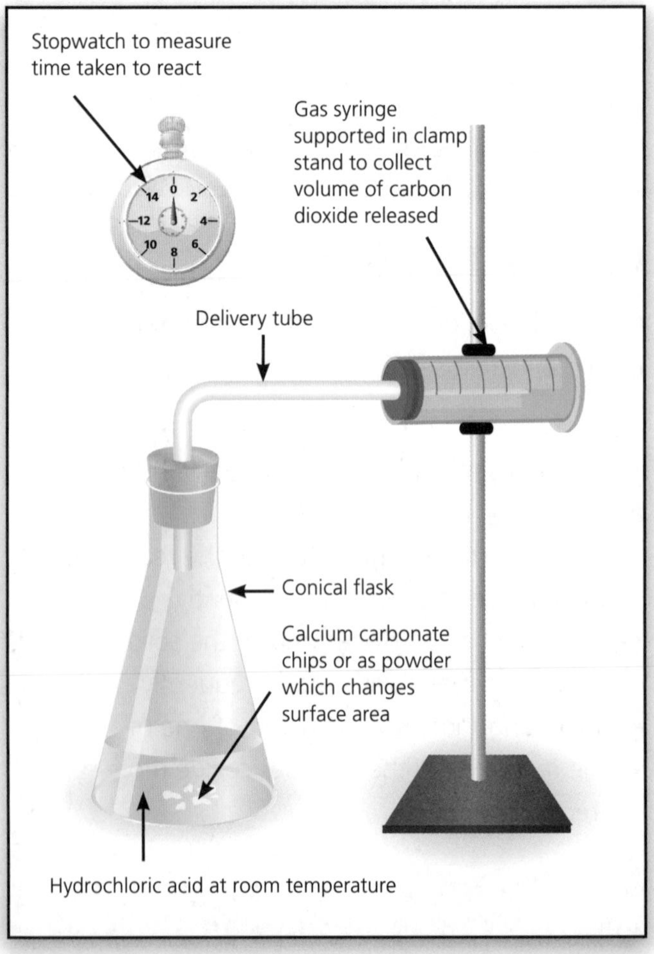
Stopwatch to measure time taken to react
Gas syringe supported in clamp stand to collect volume of carbon dioxide released
Delivery tube
Conical flask
Calcium carbonate chips or as powder which changes surface area
Hydrochloric acid at room temperature

Catalysts

A **catalyst** is a substance that increases the rate of a chemical reaction, without being used up in the process. Catalysts are often specific: different reactions require different catalysts. Because catalysts are not used up, only small amounts of them are needed.

Catalysts work by reducing the minimum energy needed for a chemical reaction to happen.

Since catalysts lower the amount of energy needed for successful collisions, higher numbers of particles will have enough energy so there will be more successful collisions per second and the reaction will be faster. Also, they can provide a surface for the molecules to attach to, thereby increasing the chances of molecules bumping into each other.

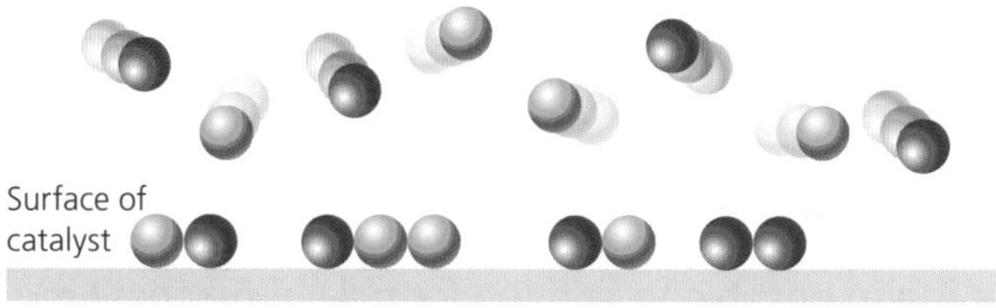
Surface of
catalyst

Below is a graph showing the decomposition of hydrogen peroxide to water, where the rate of reaction is measured by the amount of oxygen given off at one-minute intervals. The reaction is shown with and without a catalyst. As the graph shows, the reaction happens very slowly, unless a catalyst is added.

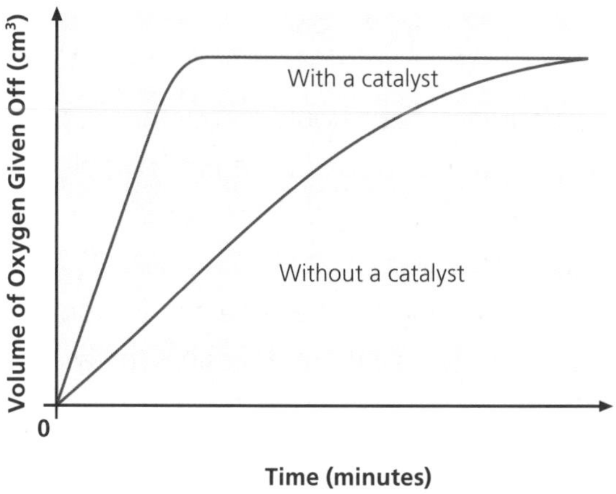
Volume of Oxygen Given Off (cm³)
With a catalyst
Without a catalyst
0
Time (minutes)

Catalytic Converters in Cars

A catalytic converter will change pollutant gases in car exhaust fumes into less harmful gases.

A catalytic converter contains a honeycomb filter where each individual cell is coated with catalysing metals, usually a mixture of platinum and rhodium. This structure increases the surface area of the converter, which increases the rate of conversion of the gas carbon monoxide into carbon dioxide.

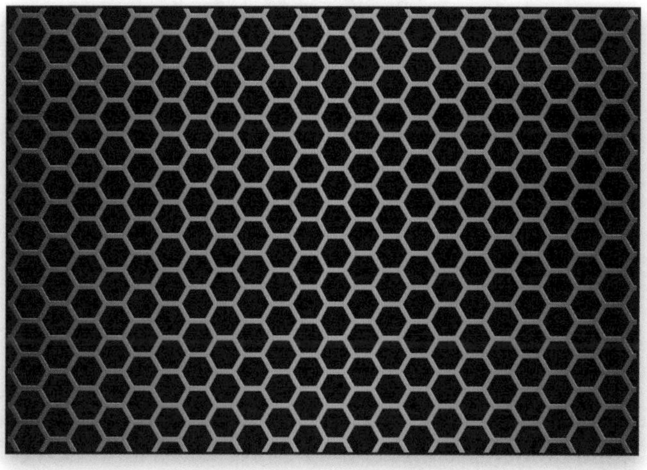

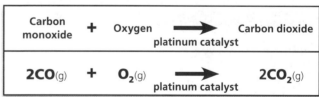

Carbon monoxide	**+**	**Oxygen**	$\longrightarrow$ **platinum catalyst**	**Carbon dioxide**
$2CO_{(g)}$	**+**	$O_{2(g)}$	$\longrightarrow$ **platinum catalyst**	$2CO_{2(g)}$

The catalytic converter also helps to oxidise unburned fuel to produce carbon dioxide and water. To help convert the gases the catalytic converter works best at a higher temperature. Increasing the temperature will also have the effect of increasing the rate of conversion of the gases.

C2 Topic 6: Quantitative Chemistry

This topic looks at:

- how different types of calculations are used
- how balanced equations are used in reacting mass calculations
- what chemical yields are
- how industry uses chemical reactions

Relative Formula Mass

The **relative formula mass** of a compound is the sum of the relative atomic masses of all its elements added together. To calculate it, we need to know the formula of the compound and the relative atomic masses of all the atoms involved.

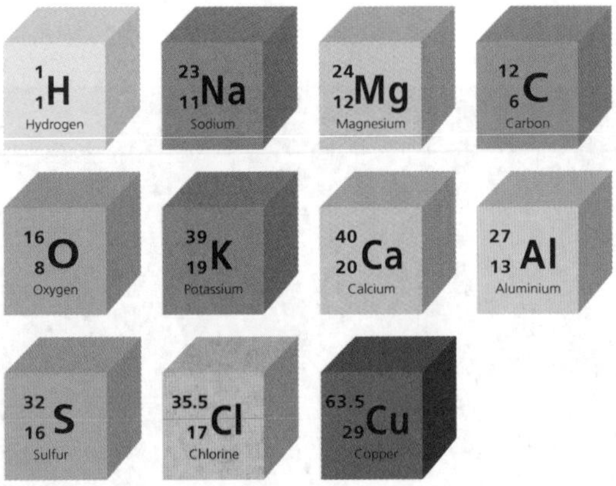

1 1 H Hydrogen
23 11 Na Sodium
24 12 Mg Magnesium
12 6 C Carbon
16 8 O Oxygen
39 19 K Potassium
40 20 Ca Calcium
27 13 Al Aluminium
32 16 S Sulfur
35.5 17 Cl Chlorine
63.5 29 Cu Copper

Example 1

Using the data above, calculate the relative formula mass of water, H_2O.

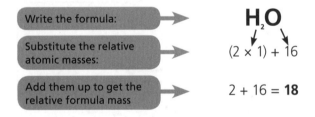

Write the formula: → H_2O

Substitute the relative atomic masses: → $(2 \times 1) + 16$

Add them up to get the relative formula mass → $2 + 16 = \mathbf{18}$

Example 2

Using the data above, calculate the relative formula mass of potassium carbonate, K_2CO_3.

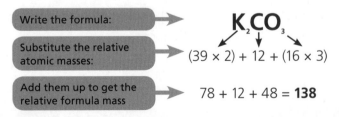

Write the formula: → K_2CO_3

Substitute the relative atomic masses: → $(39 \times 2) + 12 + (16 \times 3)$

Add them up to get the relative formula mass → $78 + 12 + 48 = \mathbf{138}$

Empirical Formula

An **empirical formula** is the simplest whole-number formula that represents the ratio of atoms in a compound. There is one simple rule: **always divide the data you are given by the relative atomic mass of the element**. Then simplify the ratio to give you the simplest formula.

Example

Find the empirical formula of an oxide of iron, produced by reacting 1.12g of iron with 0.48g of oxygen.
(Relative atomic masses: Fe = 56; O = 16)

Identify the mass of the elements in the compound:

Masses: Fe = 1.12, O = 0.48

Divide these masses by their relative atomic masses:

$Fe = \frac{1.12}{56} = 0.02 \qquad O = \frac{0.48}{16} = 0.03$

Identify the ratio of atoms in the compound and simplify it:

Ratio = 0.02 : 0.03 → × 100 (0.02 : 0.03 → 2 : 3) × 100

Empirical formula = $\mathbf{Fe_2O_3}$

It is also possible to find the formula of magnesium oxide from an experiment.

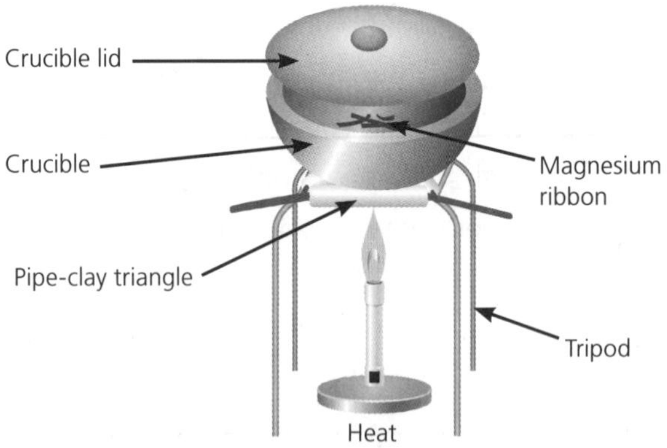

Crucible lid
Crucible
Magnesium ribbon
Pipe-clay triangle
Tripod
Heat

In this experiment, you should follow these steps.

1. First measure the mass of the empty crucible and lid.
2. Then measure the mass of the crucible with the magnesium inside.
3. Once the crucible has been heated for long enough to react all the magnesium, measure the final mass.
4. From the mass measurements, work out the empirical formula.

Percentage Composition by Mass

Using the relative formula mass you can determine the percentage composition by mass of a compound.

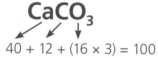

$$\text{Percentage Composition by Mass of Element} = \frac{\text{RAM of Element}}{\text{RFM of Compound}} \times 100$$

Example 1

Find the percentage composition by mass of carbon in calcium carbonate, $CaCO_3$.
(Relative atomic masses: Ca = 40; C = 12; O = 16)

1. First, work out the relative formula mass of $CaCO_3$.

$$CaCO_3$$

$$40 + 12 + (16 \times 3) = 100$$

2. Substitute the relative masses of carbon and calcium carbonate into the formula:

$$\% \text{ C} = \frac{\text{RAM of C}}{\text{RFM of } CaCO_3} \times 100 = \frac{12}{100} \times 100 = 12\%$$

Example 2

Find the percentage composition by mass of copper in copper sulfate, $CuSO_4$.
(Relative atomic masses: Cu = 63.5; S = 32; O = 16)

1. Relative formula mass of $CuSO_4$ = $63.5 + 32 + (16 \times 4) = 159.5$
2. Substituting the relative masses of Cu and $CuSO_4$ into the formula:

$$\% \text{ Cu} = \frac{63.5}{159.5} \times 100 = \mathbf{39.8\%}$$

HT Reacting Masses

All reactions follow the **law of conservation of mass**, which states that mass can neither be created nor destroyed. Therefore:

Total Mass of Reactants = Total Mass of Products

Total Mass of Reactants
Total Mass of Products

The law of conservation of mass can be used to work out how much reactant you need in a reaction, or how much product you will get from a reaction.

Calculating the Mass of a Product and a Reactant

Sometimes, we need to be able to work out how much of a substance is used up or produced in a chemical reaction.

Example 1

How much calcium oxide can be produced from 50kg of calcium carbonate? (Relative atomic masses: Ca = 40; C = 12; O = 16)

Write down the equation:

$$CaCO_3 \rightarrow CaO + CO_2$$

Work out the relative formula mass of each substance:

$$40 + 12 + (3 \times 16) \rightarrow (40 + 16) + [12 + (2 \times 16)]$$

Check the total mass of reactants equals the total mass of the products. If they are not the same, check your work:

$$100 \rightarrow 56 + 44 \checkmark$$

Since the question only mentions calcium oxide and calcium carbonate, you can now ignore the carbon dioxide. You just need the ratio of mass of $CaCO_3$ to mass of CaO.

If 100kg of $CaCO_3$ produces 56kg of CaO, then 50kg of $CaCO_3$ produces $\frac{56}{2}$ kg of CaO = **28kg** of CaO.

Calculating the Mass of a Product and a Reactant (cont.)

Example 2

How much aluminium oxide (Al_2O_3) is needed to produce 540kg of aluminium?

(Relative atomic masses: Al = 27; O = 16)

Equation:

$$2Al_2O_{3(l)} \rightarrow 4Al_{(l)} + 3O_{2(g)}$$

Masses:

$$2[(27 \times 2) + (16 \times 3)] \rightarrow (27 \times 4) + [3 \times (16 \times 2)]$$

Reactants → Products:

204 → 108 + 96

For this example we only need to look at Al_2O_3 and Al.

204kg of aluminium oxide (Al_2O_3) produces 108kg of aluminium (Al), so:

$\frac{204}{108}$kg of Al_2O_3 will produce 1kg of Al

1.89kg of Al_2O_3 will produce 1kg of Al

But we want 540kg of Al.

Therefore, the amount of Al_2O_3 needed to produce 540kg of Al is:

540 × 1.89kg = **1021kg of Al_2O_3**

Example 3

The equation for the reaction between iron and sulfur is:

$$Fe_{(s)} + S_{(s)} \rightarrow FeS_{(s)}$$

When 14g of iron is heated with excess sulfur, how much iron(II) sulfide is formed?

(Relative atomic masses: Fe = 56; S = 32)

Formula mass of FeS = 56 + 32 = 88g

From the equation, 1 atom of Fe gives 1 formula of FeS, so:

56g of Fe give 88g of FeS

1g of Fe gives $\frac{88}{56}$g of FeS

So, 14g of Fe gives 14 × $\frac{88}{56}$g of FeS

= **22g** of FeS

Example 4

The reaction between magnesium and oxygen is:

$$2Mg_{(s)} + O_{2(g)} \rightarrow 2MgO_{(s)}$$

How many grams of oxygen will react with 44g of magnesium?

(Relative atomic masses: Mg = 24; O = 16)

Formula mass of O_2 = 16 × 2 = 32g

From the equation, 2 atoms of Mg react with 1 formula (molecule) of O_2.

2 × 24g of Mg reacts with 32g of O_2

48g of Mg reacts with 32g of O_2

So, 1g of Mg reacts with $\frac{32}{48}$g of O_2

Therefore 44g of Mg reacts with 44 × $\frac{32}{48}$g of O_2

= **29.3g** of O_2

Product Yields

Chemical reactions often produce more than one **product**, and not all of these products are 'useful'. This means that not all of the starting materials (**reactants**) are converted into useful products. In a chemical reaction the amount of the useful product is known as the **yield**. The actual yield obtained from a chemical reaction is usually less than the expected yield.

The **expected or theoretical yield** is the amount of product expected from a reaction, based on the amount of reactants. Companies like to get the highest possible yield from the reaction, for the lowest cost. There are therefore two yields in a reaction:

- **theoretical yield** – calculated from the relative formula masses of reactants and products
- **actual yield** – the actual mass of useful product obtained from the reaction in the experiment.

From comparing these two yields, we can calculate the percentage yield:

$$\textbf{Percentage yield} = \frac{\textbf{Actual yield}}{\textbf{Theoretical yield}} \times 100\%$$

Example

In a chemical reaction 360g of silver nitrate, $AgNO_3$, was reacted with excess magnesium chloride, $MgCl_2$, and found to produce 264g of silver chloride, AgCl.

Calculate the percentage yield of this reaction.

(Relative atomic masses: Ag = 108; Cl = 35.5; N = 14; O = 16)

1. Write down the equation of the reaction.

Silver nitrate	+	Magnesium chloride	→	Silver chloride	+	Magnesium nitrate

$$\mathbf{2AgNO_3}_{(aq)} + \mathbf{MgCl_2}_{(aq)} \rightarrow \mathbf{2AgCl}_{(s)} + \mathbf{Mg(NO_3)_2}_{(aq)}$$

N.B. This must be the balanced equation for the reaction. You can make sure it is correct by checking that the mass of the reactants is equal to the mass of the products.

2. Work out the relative formula masses for $AgNO_3$ and AgCl using the relative atomic masses given.

   $AgNO_3 = 108 + 14 + (16 \times 3) = 170$

   $AgCl = 108 + 35.5 = 143.5$

3. The formula mass of $AgNO_3$ = 170g

   So, 360g of $AgNO_3 = \frac{360}{170} = 2.12$ formulas

4. From the equation of the reaction, 2 formulas of $AgNO_3$ produces 2 formulas of AgCl, so the ratio is 1 : 1.

   Therefore 2.12 formulas of $AgNO_3$ produces 2.12 formulas of AgCl.

   The mass of AgCl that this represents is $2.12 \times 143.5 = 304.22$g of AgCl.

   So, the theoretical yield for the reaction is 304.22g AgCl.

5. Actual yield obtained was 264g of AgCl.

   So, percentage yield = $\frac{264}{304.22} \times 100 = \mathbf{86.8\%}$.

From the example shown above, the percentage yield was not 100%, which is what an industrial process aims for. There are several factors involved in the yield of a chemical reaction:

- Chemical reactions may not be completed in the time available.
- Product may be lost during the purification procedure.
- Other reactions may be happening at the same time as the main reaction.

Not all chemical reactions produce only one product. Most produce waste. In the previous example only silver chloride was the useful product. The magnesium nitrate would have been wasted and disposed of because there was no commercial use for the material. For the company this can result in:

- an increase in process cost in order to dispose of the waste safely
- an increase in environmental concerns associated with the disposal method, as this could lead to air, water or land pollution.

Economics and the Chemical Industry

In the chemical industry chemists are constantly working to find the economically most favourable reactions. They are looking for the following key points:

- High percentage yield – a low yield will mean that more useful product would need to be produced from additional reactions.
- All products are commercially useful – if the reaction used for a particular process produces substances that have no use, then those particular products need to be disposed of, which costs money.
- Reactions occur at a suitable speed – a chemical reaction that is too slow would result in an expensive product, but a reaction that is too fast reaction could be potentially dangerous.

For example, in the Haber process (see page 71), which produces ammonia on a large scale, it is important to get the maximum yield in the shortest possible time:

- A **low temperature** increases the yield of ammonia but the reaction is too slow.
- A **high pressure** increases the yield of ammonia but the process is too expensive.
- A **catalyst** increases the rate at which equilibrium is reached but does not affect the yield of ammonia.

Any hydrogen and nitrogen gas that remains unreacted is recycled back into the main reaction vessel. The heat generated as a result of the reaction is used to heat the catalyst and the process.

Exam Practice Questions C2

Questions labelled with an asterisk (*) are ones where the quality of your written communication will be assessed – you should take particular care with your spelling, punctuation and grammar, as well as the clarity of expression, on these questions.

1 **(a)** What type of structure do metal atoms form? **(1)**

(b) What do metal structures contain that allow them to conduct electricity and heat easily? **(1)**

(c) Transition metals have properties of typical metals. List three properties you would expect a transition metal to have. **(3)**

2 **(a)** Describe the difference between theoretical yield and actual yield. **(2)**

(b) Why is it important to obtain as high a yield as possible? **(1)**

(c) The equation shows the reaction for the decomposition of zinc carbonate when it is heated.

$$\mathbf{ZnCO_3}(s) \longrightarrow \mathbf{ZnO}(s) + \mathbf{CO_2}(g)$$

(Relative atomic masses: Zn = 65; O = 16; C = 12)

(i) In an experiment 50.0g of zinc carbonate produced 31.5g of zinc oxide. The theoretical yield of zinc oxide for this reaction is 32.4g. Calculate the percentage yield. **(1)**

(ii) Calculate the percentage composition by mass of zinc in zinc oxide. **(2)**

(iii) What mass of zinc is in the 31.5g of zinc oxide formed in the experiment? **(1)**

3 **(a)** How would you show that the chemical reaction between sodium and chlorine is exothermic? **(1)**

(b) Describe how the energy involved in making and breaking bonds accounts for this type of reaction. **(3)**

(c) Which element will bromine displace, chlorine or iodine from solutions of their salts? Explain your answer. **(2)**

4 Describe why helium is used in airships and weather balloons. **(1)**

5 *Iodine has a simple molecular structure. Describe in as much detail as you can the type of bonding and resultant properties shown by iodine. **(6)**

6 Use the information below to answer the following questions.

Mass number ↓

$^{24}_{12}\mathrm{Mg}$ $^{14}_{7}\mathrm{N}$ $^{32}_{16}\mathrm{S}$

↑ Atomic number

(a) In each of these elements, describe how the number of protons would be determined. **(1)**

(b) What is the number of electrons in each element? **(3)**

(c) How is the number of neutrons in each element calculated? **(1)**

(d) Using this information, draw the electronic configuration for each element. **(3)**

7 **(a)** State three ways in which a rate of chemical reaction can be changed. **(3)**

(b) What effect does increasing the frequency of collisions does what to a rate of reaction? **(1)**

(c) Why does a low concentration of particles result in a slower rate of reaction? **(2)**

(d) Describe an experiment to show how changing the surface area of calcium carbonate when it is reacted with hydrochloric acid can produce a change in the rate of reaction. **(4)**

(e) How does using a catalyst affect a rate of chemical reaction? **(3)**

8 **(a)** Ionic bonds are formed between which two types of atom?

A ☐ metal atom and metal atom

B ☐ metal atom and noble gas

C ☐ metal atom and non-metal atom

D ☐ non-metal atom and non-metal atom **(1)**

(b) An ionic bond involves

A ☐ sharing of electrons.

B ☐ transfer of protons.

C ☐ transfer of electrons.

D ☐ sharing of protons. **(1)**

(c) Predict the formula of the following ionic compounds:

(i) Barium chloride (Ba^{2+}, Cl^{-}) **(1)**

(ii) Magnesium oxide (Mg^{2+}, O^{2-}) **(1)**

(iii) Aluminium chloride (Al^{3+}, Cl^{-}) **(1)**

(iv) Iron oxide (Fe^{3+}, O^{2-}) **(1)**

HT 9 **(a)** Write the balanced symbol equation for the following reaction. **(2)**

Magnesium + Chlorine ⟶ Magnesium chloride

(b) Write the balanced symbol equation for the following reaction. **(2)**

Iron + Oxygen ⟶ Iron(III) oxide

10 Iron is made when aluminium reacts with iron oxide. This reaction can be shown by the following balanced symbol equation.

$$\mathbf{Fe_2O_3(s) + 2Al(s) \longrightarrow 2Fe(s) + Al_2O_3(s)}$$

(Relative atomic masses: Fe = 56; Al = 27; O = 16)

(a) Work out the relative formula mass of each of the following:

(i) Fe_2O_3 **(ii)** Al_2O_3 **(2)**

(b) (i) What is the mass of aluminium needed to react with 800g of iron(III) oxide? **(3)**

(ii) What mass of iron is produced from 480g of iron(III) oxide? **(3)**

(iii) How much aluminium will be needed if 612g of aluminium oxide is produced? **(3)**

11 Diamond and graphite are two different forms of carbon. Describe the structure and properties of:

(a) diamond **(3)**

(b) graphite. **(3)**

C3 Topic 1: Qualitative Analysis

This topic looks at:

- the differences between qualitative and quantitative analysis
- why chemical tests are unique
- how scientists can use different chemical tests

Identifying Substances

In fields such as forensic science and medicine, substances need to be **identified**, and their **purity** determined.

If a substance contains a high level of **impurities**, it could be harmful. Finding out exactly which chemicals make up the substance allows scientists to find out whether they are harmful or not.

Analysis of a substance can be **qualitative** or **quantitative**.

- **Qualitative analysis** – any method used to identify **which chemicals are present**, for example, using an indicator to find out if acids are present.
- **Quantitative analysis** – any method used to determine the **amount of a particular chemical present**, for example, carrying out an acid–base titration to find out how much acid is present.

Identifying Ions

An ionic substance can be identified by determining each type of **ion** within it. Each ion has its own unique test so it is easy to identify.

Metal compounds in solution contain either:

- **metal ions** and **non-metal ions**, or
- **metal ions** and **compound ions**

1. First identify the metal ion by using flame tests or precipitate tests.
2. Then identify the non-metal or compound ions.

Flame Tests for Metal Ions

1. A piece of nichrome (a nickel-chromium alloy) wire is dipped in concentrated hydrochloric acid and held in a blue Bunsen flame to clean it. This is repeated until the flame around the wire burns clean.

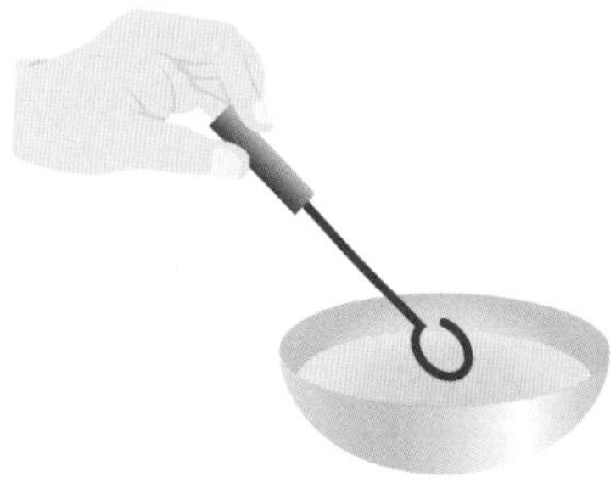

2. The compound to be tested is made into a paste using a small amount of hydrochloric acid. The nichrome wire is then dipped into the compound and a small sample collected.

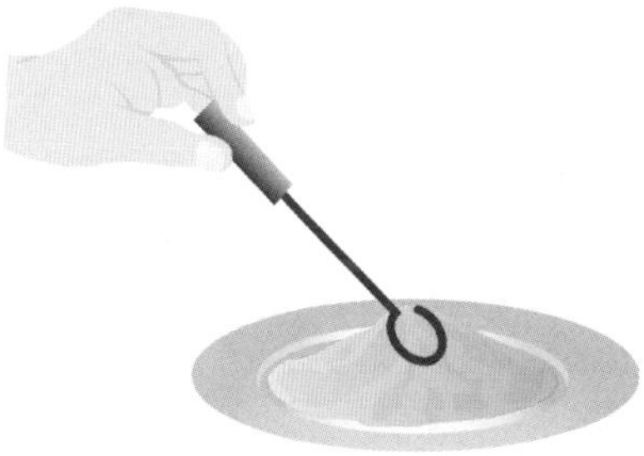

3. The sample on the wire is then put into the blue Bunsen flame, which should burn with a distinctive colour. The distinctive colour produced can be used to identify the metal ion.

Sodium (Yellow)

Potassium (Lilac)

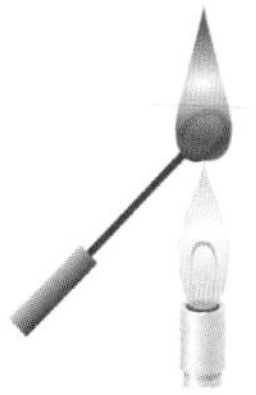

Calcium (Brick red)

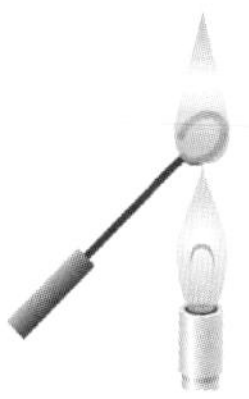

Copper (Blue-green)

C3 Qualitative Analysis

Precipitate Tests for Metal Ions

Some metal compounds form precipitates when specific substances are added to them. To test for metal ions, sodium hydroxide solution is added to a solution of an ionic substance. If the ionic substance contains **metal ions** then the following coloured precipitates will be produced.

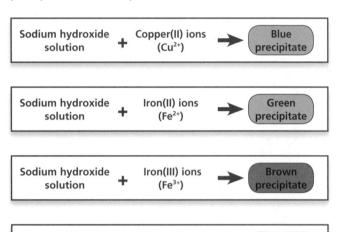

Sodium hydroxide solution	+	Copper(II) ions (Cu^{2+})	→	Blue precipitate

Sodium hydroxide solution	+	Iron(II) ions (Fe^{2+})	→	Green precipitate

Sodium hydroxide solution	+	Iron(III) ions (Fe^{3+})	→	Brown precipitate

Sodium hydroxide solution	+	Aluminium ions (Al^{3+})	→	White precipitate

Calcium ions (Ca^{2+}) also form white precipitates when added to sodium hydroxide solution.

Calcium ions form **calcium hydroxide**.

$$Ca^{2+}{}_{(aq)} + 2OH^-{}_{(aq)} \rightarrow Ca(OH)_{2(s)}$$

Aluminium ions form **aluminium hydroxide**, which dissolves in excess sodium hydroxide to form sodium aluminate $NaAlO_2$.

$$Al^{3+}{}_{(aq)} + 3OH^-{}_{(aq)} \rightarrow Al(OH)_{3(s)}$$

Test for Halide Ions

To test for halide ions, dilute nitric acid and silver nitrate solution are added to a solution of an ionic substance. If the ionic substance contains **halide ions** (i.e. chloride, bromide or iodide) then the following coloured precipitates will be produced.

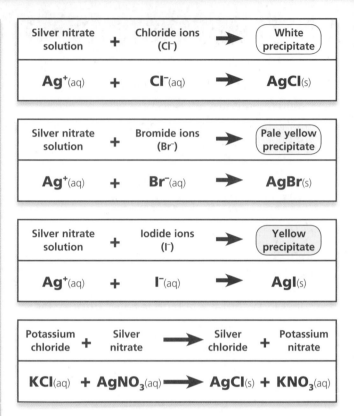

Silver nitrate solution	+	Chloride ions (Cl^-)	→	White precipitate
$Ag^+{}_{(aq)}$	+	$Cl^-{}_{(aq)}$	→	$AgCl_{(s)}$

Silver nitrate solution	+	Bromide ions (Br^-)	→	Pale yellow precipitate
$Ag^+{}_{(aq)}$	+	$Br^-{}_{(aq)}$	→	$AgBr_{(s)}$

Silver nitrate solution	+	Iodide ions (I^-)	→	Yellow precipitate
$Ag^+{}_{(aq)}$	+	$I^-{}_{(aq)}$	→	$AgI_{(s)}$

Potassium chloride	+	Silver nitrate	→	Silver chloride	+	Potassium nitrate
$KCl_{(aq)}$	+	$AgNO_{3(aq)}$	→	$AgCl_{(s)}$	+	$KNO_{3(aq)}$

Identifying Ammonium Ions (NH_4^+)

When sodium hydroxide solution (which contains OH^- ions) is added to **NH_4^+ ions** in solution and heated, **ammonia gas** is given off.

Therefore, if sodium hydroxide solution is added to an unknown substance and it gives off ammonia gas when heated, the substance contains NH_4^+ ions. The reaction also produces water.

Ammonium ions	+	Hydroxide ions	→	Ammonia	+	Water
$NH_4^+{}_{(aq)}$	+	$OH^-{}_{(aq)}$	→	$NH_{3(g)}$	+	$H_2O_{(l)}$

We can test for ammonia gas using damp red litmus paper. If ammonia is present, the damp red litmus paper turns blue.

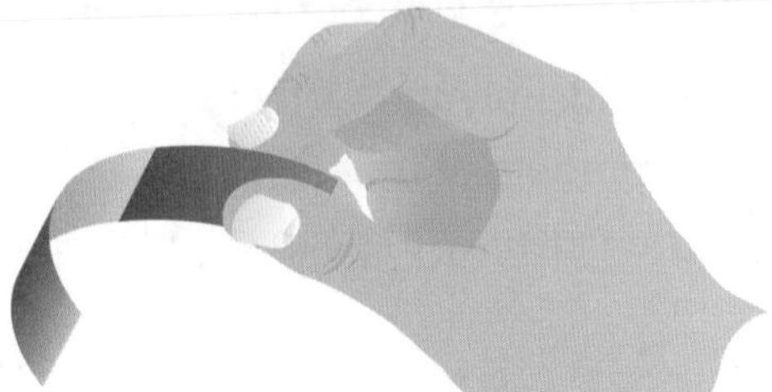

Tests for Ions in Industry

Many industries use tests for ions, including:
- to check the purity of water
- to check medical patients for mineral deficiencies.

Water Purity

Water purity needs to carefully monitored, because:
- hard water will produce high levels of scum, which can impact on industrial and domestic machinery, such as washing machines
- acidic water could result in high levels of gastrointestinal-related illnesses if used for drinking water
- if water is polluted with halogens it could interfere with some chemical reactions when used in industry.

The hardness of water is checked by determining which impurities are present (qualitative analysis), for example, testing for magnesium ions, calcium ions, sodium ions and carbonate ions, and how much of each impurity is present (quantitative analysis). A decision can then be made as to whether the compounds present will interfere with any processes for which the water will be used and the water can be softened if necessary.

The acidity of water is checked by testing the pH level.

Testing for chloride and bromide ions allows scientists to check if water contains compounds of chlorine and bromine.

Mineral Deficiencies

Minerals in the diet are very important for health. Simple ion tests can be carried out on blood to check for a mineral deficiency. A device called a flame photometer is used for the flame tests.

Mineral	Function	Test
Calcium	• Healthy bones • Muscle contractions • Brain functions • Blood clotting	Flame test
Potassium	• Regulates blood pressure • Healthy heart	Flame test
Copper	• Growth and development	Flame test
Sodium	• Transmitting electrical (nervous) impulses	Flame test
Chlorides	• Fluid balance	Halide testing
Iodides	• Thyroid functions	Halide testing

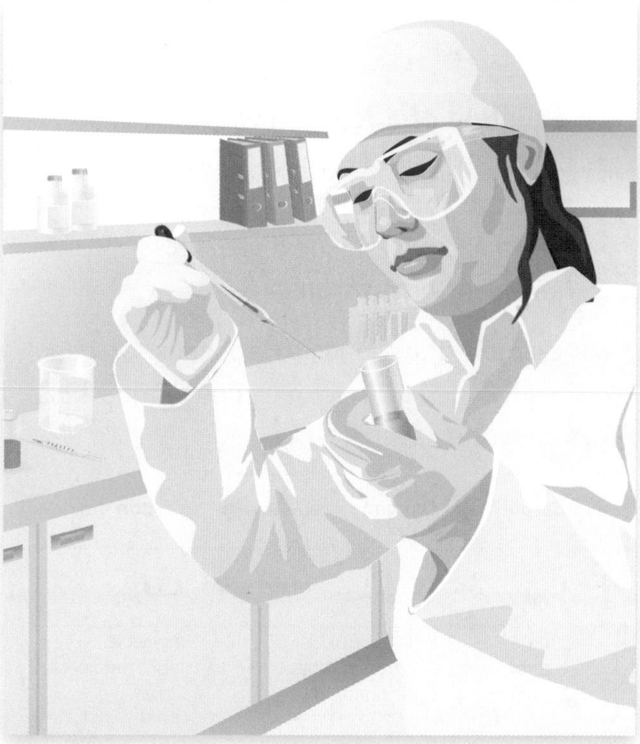

C3 Quantitative Analysis

C3 Topic 2: Quantitative Analysis

This topic looks at:
- how mole, mass and concentration calculations can be used
- how to calculate solution concentrations
- how titrations can be used
- the different methods of preparing soluble salts
- what makes water hard and the associated problems

Quantitative Analysis

Quantitative analysis is calculating **how much** there is of a particular chemical in a substance.

Chemists use **formulae** and **balanced symbol equations** to help them find out:
- what they have produced
- how much of it they have produced
- how product-efficient a reaction is.

Decisions can then be made as to whether the reactions need to be researched further.

HT The amount of a substance can be measured in **grams, number of particles** or **number of moles of particles**.

Equations

An **equation of a reaction** will tell you:
- how many particles of reactants and particles there are
- how many moles (mol) there are
- how many grams there are.

For example, the elements carbon and oxygen will react together to form carbon dioxide. This can be written as follows.

Carbon	+ Oxygen	→ Carbon dioxide
$C_{(g)}$	+ $O_{2(g)}$	→ $CO_{2(g)}$
1 atom	+ 1 molecule	1 molecule
1 mol of atoms	+ 1 mol of molecules	1 mol of molecules
12 grams	+ 32 grams	44 grams

HT Moles

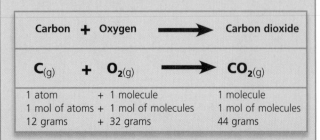

A **mole** is a measure of the amount of substance. A mole is a very large number of particles. A mole of carbon atoms contains the **same number of particles** as a mole of sodium atoms or a mole of carbon dioxide molecules.

The mass of a mole of **atoms** of any **element** is always equal to the **relative atomic mass** of the element, expressed in grams per mole.

For substances that are **molecules** or **compounds**, the mass of a mole of the substance is equal to the **relative formula mass** (i.e. the relative atomic masses of all its component elements added together) of the substance, expressed in grams per mole.

Converting Moles to Masses

It is easy to convert moles to masses if you use **relative atomic mass** (see pages 27 and 28) and **relative formula mass** (see page 48).

The relative atomic mass is the number at the top left of the element symbol in the periodic table.

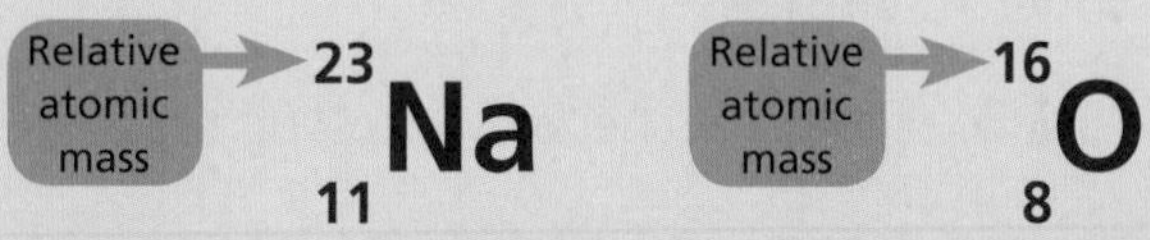

Relative atomic mass
23
Na
11
Relative atomic mass
16
O
8

For sodium, the relative atomic mass is 23. This means a mole of sodium atoms has a mass of 23g. The atomic mass of sodium is 23g mol^{-1}.

Similarly, a mole of oxygen *atoms* has a mass of 16g, since the relative atomic mass of oxygen is 16. The atomic mass of oxygen is 16g mol^{-1}.

But the element oxygen exists as a *molecule*, O_2. The relative formula mass of O_2 is (2 × 16) = 32, so a mole of oxygen *molecules* has a mass of 32g. The formula mass of oxygen is 32g mol^{-1}.

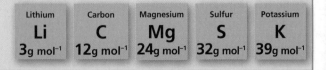

Lithium	Carbon	Magnesium	Sulfur	Potassium
Li	C	Mg	S	K
3g mol^{-1}	12g mol^{-1}	24g mol^{-1}	32g mol^{-1}	39g mol^{-1}

Calculations Using Moles

HT

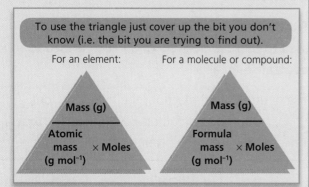

To use the triangle just cover up the bit you don't know (i.e. the bit you are trying to find out).

For an element:

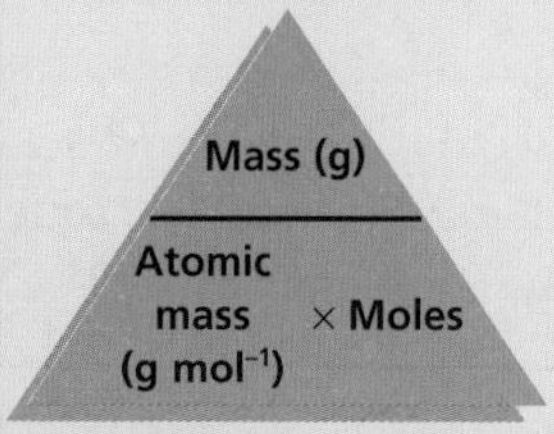
Mass (g)
Atomic mass (g mol⁻¹)
× Moles

For a molecule or compound:

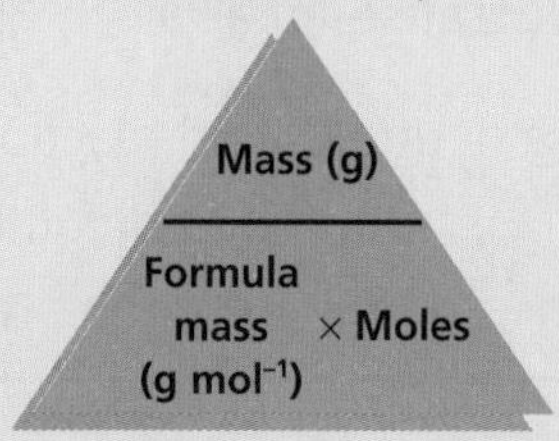
Mass (g)
Formula mass (g mol⁻¹)
× Moles

Example 1

Calculate the number of moles of carbon dioxide, CO_2, in 33g of the gas.

Relative atomic mass of C = 12
Relative atomic mass of O = 16
Relative formula mass of CO_2 = 12 + (16 × 2)
= 44

So, formula mass of CO_2 = 44g mol^{-1}
Using the triangle:

$$\text{No. of moles (mol)} = \frac{33g}{44g\ mol^{-1}}$$

= 0.75mol

Example 2

Calculate the mass of 4mol of sodium hydroxide, NaOH.

Relative atomic mass of Na = 23
Relative atomic mass of O = 16
Relative atomic mass of H = 1
Relative formula mass of NaOH = 23 + 16 + 1
= 40

So, formula mass of NaOH = 40g mol^{-1}
Using the triangle:
Mass (g) = 40g mol^{-1} × 4mol = **160g**

Example 3

Calculate the mass of 9 mol of water, H_2O.

Relative atomic mass of H = 1
Relative atomic mass of O = 16
Relative formula mass of H_2O = (1 × 2) + 16
= 18

So, formula mass of H_2O = 18g mol^{-1}
Using the triangle:
Mass (g) = 18g mol^{-1} × 9mol = **162g**

Concentrations of Solutions

Concentration is a measure of the amount of solute in 1 cubic decimetre of solution. Concentration can be measured in grams per cubic decimetre (g dm^{-3}).

Remember that 1 cubic decimetre (1dm^3) is the same as 1000 cubic centimetres (1000cm^3).

Example 1

In 250cm^3 of solution there are 6.2g of sodium chloride. What is the concentration in g dm^{-3}?

250cm^3 is $\frac{1}{4}$ of 1dm^3. So in 1dm^3 there are 4 × 6.2g of NaCl = 24.8g

So, the concentration of the NaCl solution = **24.8g dm^{-3}**.

HT

Concentration can also be expressed in moles per cubic decimetre (mol dm^{-3}).

If the concentration of a solution is:

- **1mol dm^{-3}**, then 1mol of solute is present in every 1dm^3 of solution
- **0.5mol dm^{-3}**, then 0.5mol of solute is present in every 1dm^3 of solution.

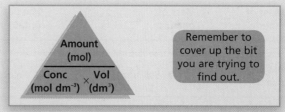

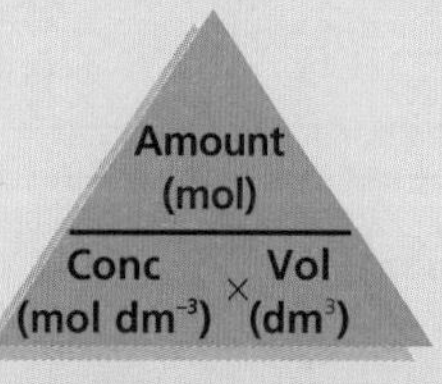
Amount (mol)
Conc (mol dm⁻³)
×
Vol (dm³)

Remember to cover up the bit you are trying to find out.

Example 2

An aqueous solution of sodium hydroxide has a volume of 25cm^3 and a concentration of 0.5mol dm^{-3}.

Calculate the number of moles of NaOH present in the solution.

Using the formula triangle:

Amount = Concentration × Volume

So, $\frac{25cm^3}{1000}$ = 0.025dm^3

Remember to convert volumes to dm^3.

Amount = 0.5mol dm^{-3} × 0.025dm^3

So, the amount of NaOH present in the solution **= 0.0125mol**

Measuring Mass Using Evaporation

The mass of a **solute** can also be measured by evaporating all the water and measuring the mass of the solid left behind.

1. Measure the mass of the empty container.
2. Measure the mass of the container and the solution.
3. Heat the solution to evaporate the solvent.
4. Measure the mass of the container and the remaining solute.
5. Gently re-heat to make sure all the solvent has been evaporated.
6. Measure the mass of the container and the solute again.

If there is no further difference in the masses, the mass of solute dissolved in the solvent has been determined.

The mass of **solvent** used has also been determined.

Mass of solute = (**Mass of container** + **Solute**) – **Mass of container**

Mass of solvent = (**Mass of container** + **Solution**) – (**Mass of container** + **Solute**)

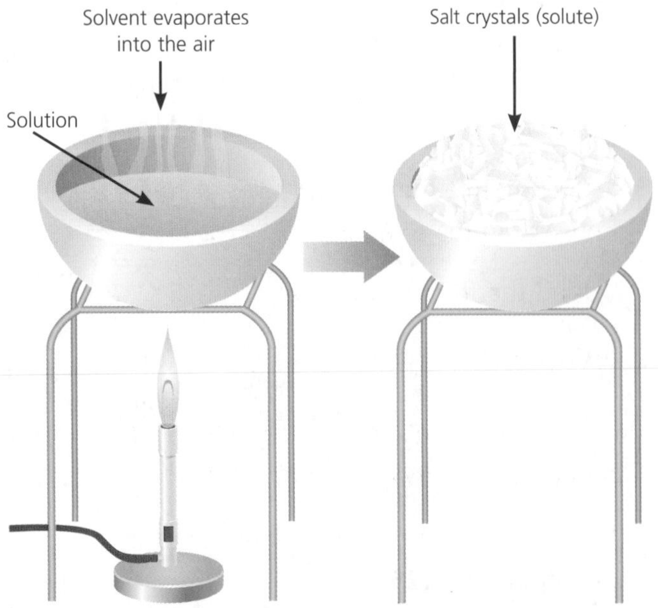
Solvent evaporates into the air
Salt crystals (solute)
Solution

Using Titrations to Standardise Solutions

Titration is an accurate technique that can be used to find the quantity of an acid needed to **neutralise** an alkali.

In a neutralisation reaction, the hydrogen ions, $H^+(aq)$, from the acid combine with the hydroxide ions, $OH^-(aq)$, from the alkali.

$$H^+(aq) + OH^-(aq) \rightarrow H_2O(l)$$

Using a pipette, an accurate volume of alkali, usually 25cm^3, is measured into a clean, dry conical flask and a few drops of indicator (e.g. phenolphthalein, which is pink in alkali) is added. The acid is added to the alkali from a burette, a little at a time. The flask is swirled in a controlled way to allow the acid and the alkali to mix.

When all the alkali has been neutralised, the indicator suddenly turns colourless, showing that the solution is neutral and no more acid needs to be added. The scale on the burette is used to determine the volume of acid added. To make a pure salt, the reaction is carried out again, but this time without the indicator.

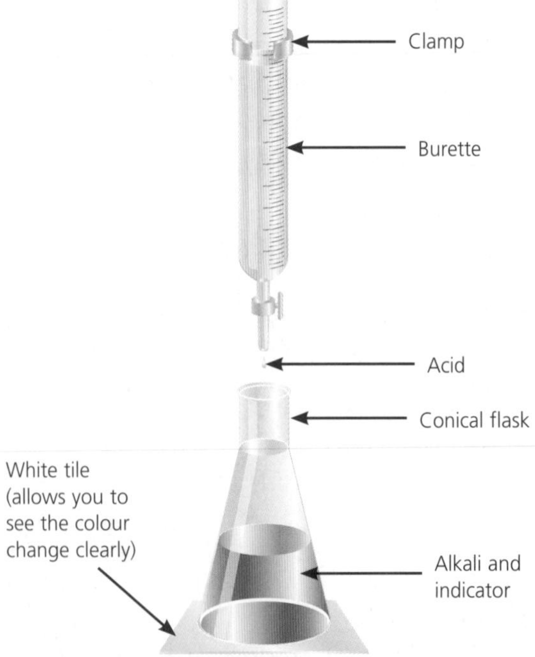
Clamp
Burette
Acid
Conical flask
White tile (allows you to see the colour change clearly)
Alkali and indicator

Remember: acid (aq) + base (aq) → salt (aq) + water (l)

HT

Calculating Concentrations from Titrations

In a neutralisation reaction, it is very easy to calculate the exact concentration of one of the solutions if the concentration of the other is known.

Example
A titration is carried out and $0.035dm^3$ of $0.6mol\ dm^{-3}$ sulfuric acid neutralises $0.14dm^3$ of sodium hydroxide solution.

a) Calculate the concentration of the sodium hydroxide solution.

$$H_2SO_4(aq) + 2NaOH(aq) \rightarrow Na_2SO_4(aq) + 2H_2O(l)$$

1mol of H_2SO_4 neutralises 2mol of NaOH (because only one molecule of H_2SO_4 is needed to neutralise two molecules of NaOH, as shown in the balanced symbol equation above).

Amount of sulfuric acid = Concentration × Volume
= $0.6mol\ dm^{-3} \times 0.035dm^3$
= 0.021mol

Therefore, the amount of NaOH that reacted with the sulfuric acid

= 0.021mol × 2 (because 1mol of H_2SO_4 neutralises 2mol of NaOH)

= 0.042mol

Concentration of NaOH solution $= \frac{\text{Amount}}{\text{Volume}}$

$= \frac{0.042}{0.14} =$ **$0.3mol\ dm^{-3}$**

b) Calculate the concentration in $g\ dm^{-3}$ of the sodium hydroxide solution used for this titration. (Relative atomic masses: Na = 23; O = 16; H = 1)

Relative formula mass of NaOH
= 23 + 16 + 1 = 40

So, formula mass of NaOH = $40g\ mol^{-1}$

From part **(a)**, concentration = $0.3mol\ dm^{-3}$

So, converting to $g\ dm^{-3}$,

Concentration is $(0.3 \times 40)g\ dm^{-3}$
= **$12g\ dm^{-3}$**

Titration Curves and Indicators

Indicators are dissolved dyes that can be used to show when neutralisation occurs, but no one indicator is sufficient to show the point of neutralisation of all acid–base titrations.

A titration curve is produced by plotting pH (vertical axis) against volume of acid (horizontal axis). It produces a curve of the following shape.

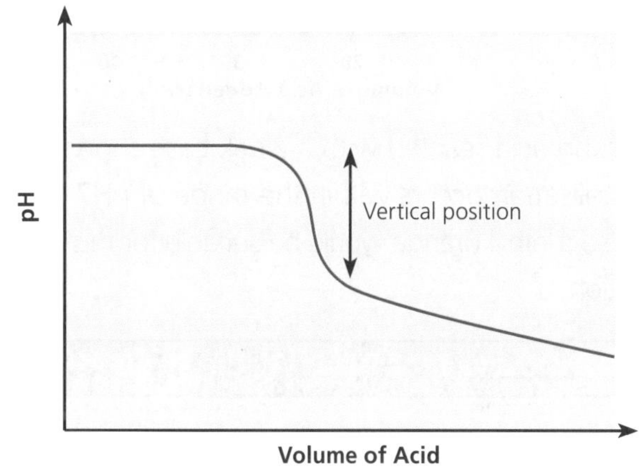
pH
Vertical position
Volume of Acid

Using titration curves helps to determine the most suitable indicator for the titration being undertaken, because the indicator needs to change colour within the vertical section of the curve.

An indicator will change colour within a range of pH values. Examples are shown in the table.

Indicator	Effective pH Range	Colour in Acid	Colour in Alkali
Methyl orange	2.1 to 4.4	Red	Yellow
Thymol blue	7.9 to 9.4	Yellow	Blue
Phenolphthalein	8.3 to 10.0	Colourless	Pink
Methyl red	4.2 to 6.2	Red	Yellow

Titration Curves and Indicators (cont.)

There are four types of titration curve, depending on the strength of the acid and base being used.

Titration Curves for Strong Acid – Weak Base

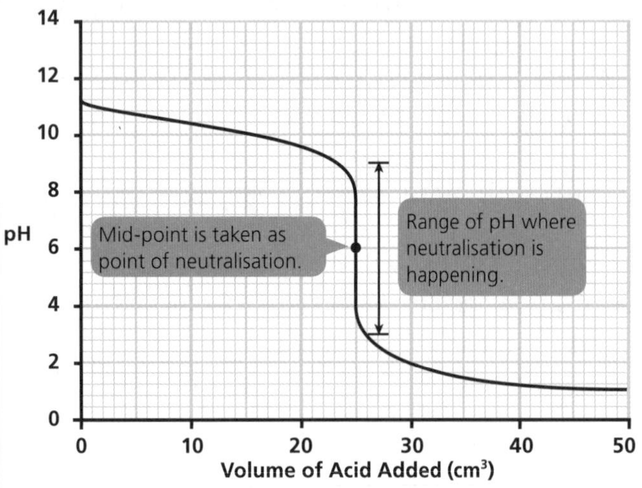
14
12
10
8
6
4
2
0
pH
Mid-point is taken as point of neutralisation.
Range of pH where neutralisation is happening.
0
10
20
30
40
50
Volume of Acid Added (cm³)

A strong acid reacted with a weak base shows that neutralisation occurs within the range of pH7 to pH3 so methyl orange would be suitable for this reaction.

Titration Curves for Strong Acid – Strong Base

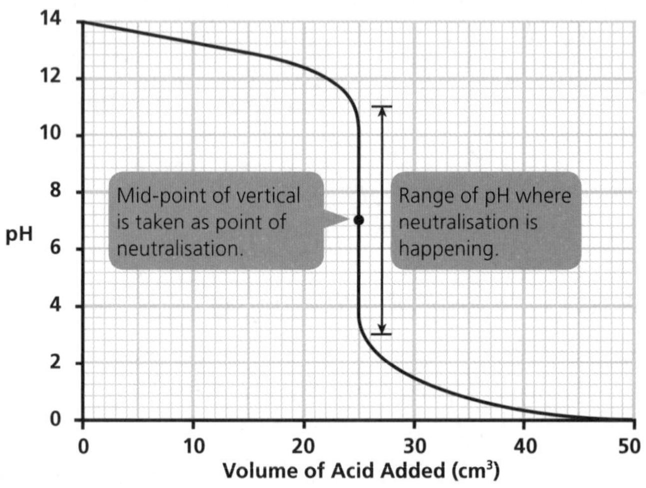
14
12
10
8
6
4
2
0
pH
Mid-point of vertical is taken as point of neutralisation.
Range of pH where neutralisation is happening.
0
10
20
30
40
50
Volume of Acid Added (cm³)

A strong acid reacted with a strong base shows that neutralisation occurs within the range of pH11 to pH3 so phenolphthalein and methyl orange would be suitable for this reaction.

Titration Curves for Weak Acid – Strong Base

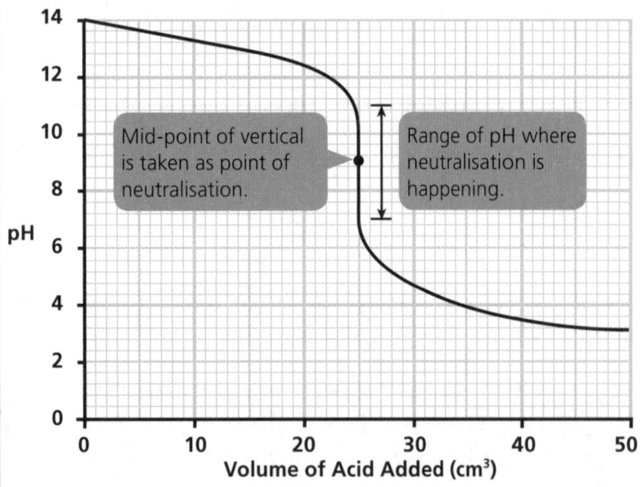
14
12
10
8
6
4
2
0
pH
Mid-point of vertical is taken as point of neutralisation.
Range of pH where neutralisation is happening.
0
10
20
30
40
50
Volume of Acid Added (cm³)

A weak acid reacted with a strong base shows that neutralisation occurs within the range of pH11 to pH7 so phenolphthalein or thymol blue would be suitable for this reaction.

Titration Curves for Weak Acid – Weak Base

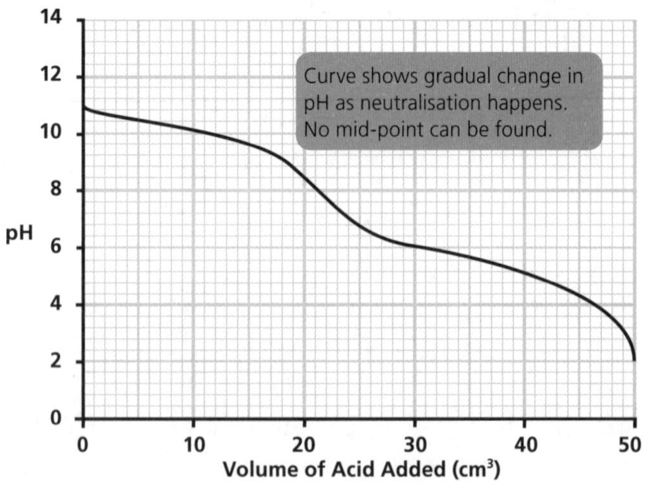
14
12
10
8
6
4
2
0
pH
Curve shows gradual change in pH as neutralisation happens. No mid-point can be found.
0
10
20
30
40
50
Volume of Acid Added (cm³)

A weak acid reacted with a weak base shows that neutralisation point is impossible to determine accurately.

Neutralisation Involving Ions

Alkalis and insoluble bases both react with acid in neutralisation reactions. For example, hydrochloric acid solution and sodium hydroxide solution react according to the following equation.

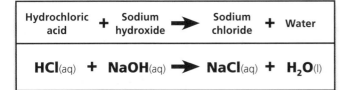

Hydrochloric acid	+	Sodium hydroxide	→	Sodium chloride	+	Water
$HCl_{(aq)}$	+	$NaOH_{(aq)}$	→	$NaCl_{(aq)}$	+	$H_2O_{(l)}$

A neutralisation reaction between any aqueous acid–base combination can be described in terms of the hydrogen ions, $H^+_{(aq)}$, from the acid and the hydroxide ions, $OH^-_{(aq)}$, from the base.

$$H^+_{(aq)} + OH^-_{(aq)} \longrightarrow H_2O_{(l)}$$

It is the concentration of the hydrogen ions in the solution that determines the strength of the acid or alkali.

Salts from Insoluble Bases

When a soluble salt is prepared from an acid and an insoluble base, then this usually means that an acid is reacted with a solid such as copper oxide. This is because the base does not dissolve in water.

For example, when making copper sulfate it is important to add excess copper oxide to the measured volume of sulfuric acid. The copper oxide is added to sulfuric acid until there is no longer a reaction and there is evidence of copper oxide left behind. To remove the excess copper oxide or residue, the solution is then filtered.

Sulfuric acid	+	Copper oxide	→	Copper sulfate	+	Water
$H_2SO_{4(aq)}$	+	$CuO_{(s)}$	→	$CuSO_{4(aq)}$	+	$H_2O_{(l)}$

Insoluble Base in Excess

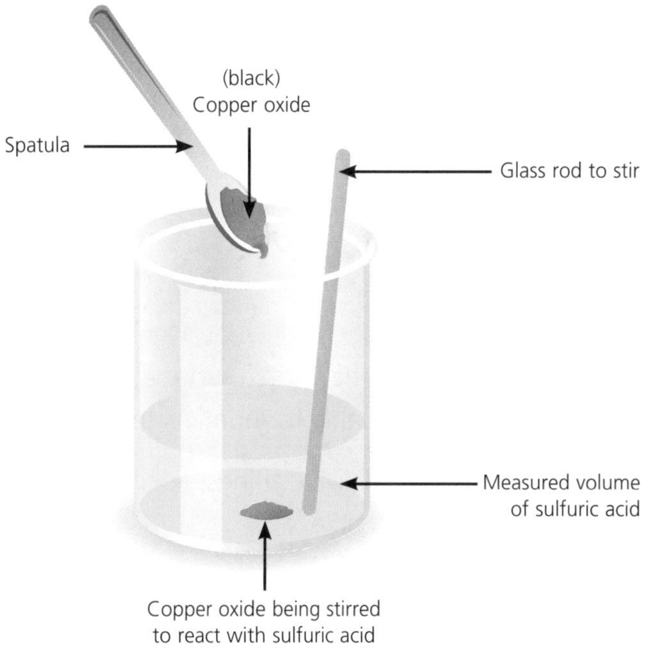
(black)
Copper oxide
Spatula
Glass rod to stir
Measured volume of sulfuric acid
Copper oxide being stirred to react with sulfuric acid

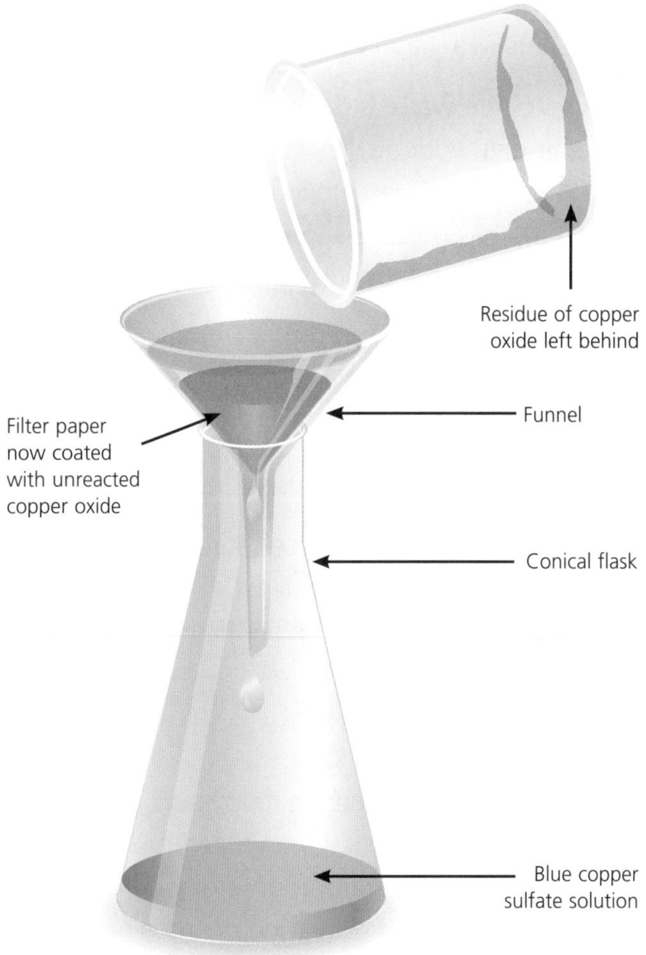
Residue of copper oxide left behind
Funnel
Filter paper now coated with unreacted copper oxide
Conical flask
Blue copper sulfate solution

Water

Water is a compound of **hydrogen** and **oxygen**. It has the formula H_2O.

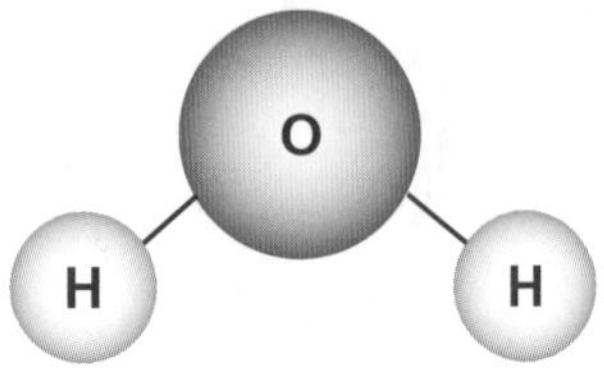
O
H
H

Water covers most of the Earth's surface, but less than 0.1% of it is available for humans to use.

Water is essential to all living things, as they all contain water, for example:
- a lettuce is about 96% water
- a human is over 70% water.

The four main sources of fresh water are:
- rivers
- lakes
- reservoirs
- aquifers (wells and boreholes).

Adequate safe water is the key to good health and a proper diet. Poor quality water and bad sanitation can be fatal. However, around 20% of the world's population has no access to safe drinking water and around 50% of the population lacks safe sanitation.

Most tap water in Britain comes from rivers and surface reservoirs.

Some tap water is pumped up through boreholes from natural underground reservoirs.

However, water from all these sources is not completely pure. It may contain:
- **bacteria** – most of which are harmless, but some can cause disease
- **dissolved substances** such as calcium compounds from rocks
- **solid substances** such as mud, sand, twigs and dumped rubbish
- **nitrates** from the run-off of fertilisers
- **pesticides** from the spraying of crops near to the water supply.

Water that is used for drinking has most of the bacteria killed and the solid substances removed at water treatment plants.

Hard and Soft Water

Water is a solvent and many compounds can dissolve in it. The hardness of water is determined by the amount of calcium ions (Ca^{2+}) and / or magnesium ions (Mg^{2+}) that are dissolved in it.

Most hard water contains calcium or magnesium compounds, which dissolve in water that flows over ground or rocks containing compounds of these elements.

Hardness in water can be temporary or permanent. Temporary hardness is caused by calcium hydrogencarbonate and / or magnesium hydrogencarbonate. This can be removed by boiling, which leaves behind limescale.

Permanent hardness is caused by calcium sulfate and magnesium sulfate salts. Permanent hardness can be removed using ion exchange resins. Water is passed through the resin, which swaps the calcium and magnesium ions for sodium ions.

Soft water does not contain many dissolved calcium or magnesium compounds.

When soap is added to hard water, more is usually needed to form a lather than would be required with soft water. This is because:
- the calcium and / or magnesium ions in hard water react with the soap
- some of the soap is used up in the reaction and forms insoluble salts
- these insoluble salts appear as a 'scum', rather than forming a lather.

C3 Topic 3: Electrolytic Processes

This topic looks at:

- electrolytes
- reactions in electrolysis and half equations
- the products from electrolysis of sodium chloride and their uses
- the products from electrolysis of aqueous solutions
- other everyday industrial uses of electrolysis

Electrolysis

When electricity is passed through an ionic substance that has been dissolved in water or a molten ionic substance, the ionic substance is broken down. This process is called **electrolysis**.

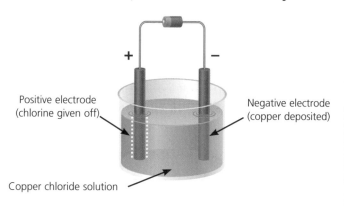

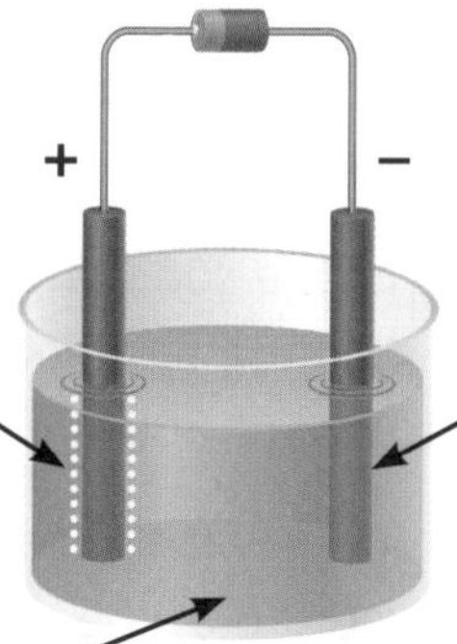
+
–
Positive electrode
(chlorine given off)
Negative electrode
(copper deposited)
Copper chloride solution

During electrolysis, charged ions move to the electrode of opposite charge.

- Negatively charged ions (**anions**) are attracted to the positively charged anode where they **lose electrons** to become atoms. This is called **oxidation**.
- Positively charged ions (**cations**) are attracted to the negatively charged cathode where they **gain electrons** to become atoms. This is called **reduction**.

Products of Electrolysis

An **electrolyte** is a substance that conducts an electric current when it is in solution or liquid form. For example, copper chloride solution is an electrolyte.

These processes usually use inert electrodes. Different electrolytes produce different substances at the electrodes (see table below). Remember that electrolysis will only work if a simple **circuit** is set up that allows the free movement of the positive and negative ions to complete the circuit.

Electrolyte	At Cathode (–)	At Anode (+)
Copper chloride solution	Copper metal	Chlorine gas
Copper sulfate solution	Copper metal	Oxygen gas
Sodium sulfate solution	Hydrogen gas	Oxygen gas
Molten lead bromide	Lead metal	Bromine gas

HT Half equations can be used to describe what happens at the electrodes by showing what happens to the ions during electrolysis.

Electrolyte	At Anode (positive electrode)	At Cathode (negative electrode)
$NaCl_{(l)}$ (Molten sodium chloride)	$2Cl^-_{(l)} \longrightarrow Cl_{2(g)} + 2e^-$	$Na^+_{(l)} + e^- \longrightarrow Na_{(l)}$
$NaCl_{(aq)}$ (Aqueous sodium chloride)	$2Cl^-_{(aq)} \longrightarrow Cl_{2(g)} + 2e^-$	$2H^+_{(aq)} + 2e^- \longrightarrow H_{2(g)}$
$PbBr_{2(l)}$ (Molten lead bromide)	$2Br^-_{(l)} \longrightarrow Br_{2(g)} + 2e^-$	$Pb^{2+}_{(l)} + 2e^- \longrightarrow Pb_{(l)}$
$CuSO_{4(aq)}$ (Aqueous copper sulfate)	$4OH^-_{(aq)} \longrightarrow 2H_2O_{(l)} + O_{2(g)} + 4e^-$	$Cu^{2+}_{(aq)} + 2e^- \longrightarrow Cu_{(s)}$
$Na_2SO_{4(aq)}$ (Aqueous sodium sulfate)	$4OH^-_{(aq)} \longrightarrow 2H_2O_{(l)} + O_{2(g)} + 4e^-$	$2H^+_{(aq)} + 2e^- \longrightarrow H_{2(g)}$

Electrolysis of Sodium Chloride

A solution of sodium chloride dissolved in water contains sodium ions (Na^+) and chloride ions (Cl^-), hydrogen ions (H^+) and hydroxide ions (OH^-). The H^+ ions and the OH^- ions are produced by water. During electrolysis the hydrogen ions and hydroxide ions will compete with the sodium ions and the chloride ions to gain or lose electrons.

The diagram shows how a simple cell can be used to investigate electrolysis of sodium chloride solution in the laboratory.

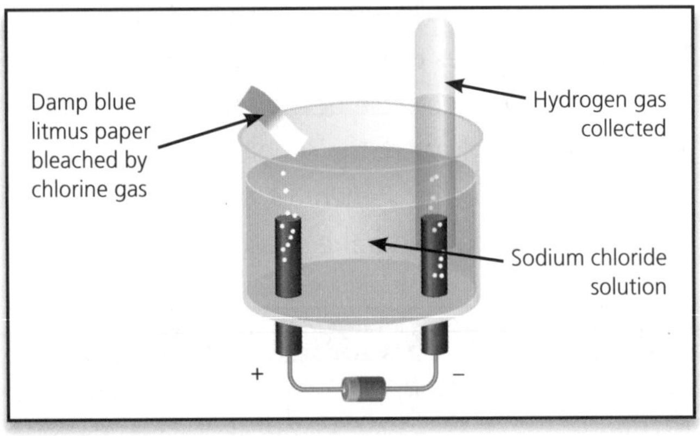
Damp blue litmus paper bleached by chlorine gas
Hydrogen gas collected
Sodium chloride solution
+
–

When the solution of sodium chloride is electrolysed, the following occurs:

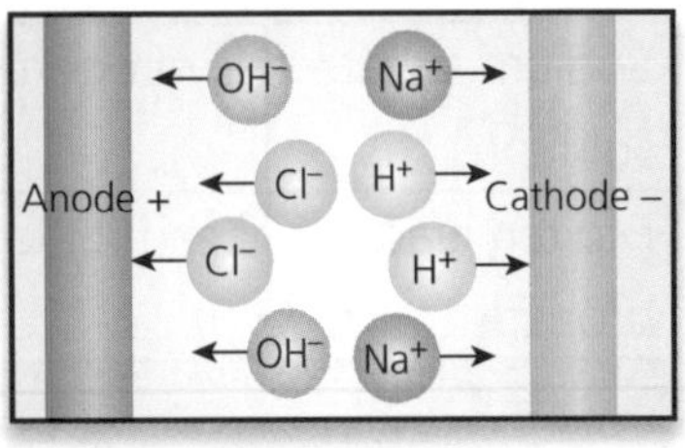
OH⁻
Na⁺
Anode +
Cl⁻
H⁺
Cathode –
Cl⁻
H⁺
OH⁻
Na⁺

- The sodium ions (Na^+) and hydrogen ions (H^+) move towards the cathode. The chloride ions (Cl^-) and hydroxide ions (OH^-) move towards the anode.
- At the cathode, the hydrogen ions (H^+) gain electrons and hydrogen gas bubbles off whilst the sodium ions remain in solution.

$$2H^+(aq) + 2e^- \longrightarrow H_2(g)$$

- At the anode the chloride ions (Cl^-) lose their electrons more easily than the hydroxide ions (OH^-), so chlorine gas bubbles off whilst the hydroxide ions remain in solution.

$$2Cl^-(aq) \longrightarrow Cl_2(g) + 2e^-$$

When the chlorine gas and the hydrogen gas bubble off the sodium ions and hydroxide ions remain and a solution of sodium hydroxide, $NaOH(aq)$, is formed.

Sodium Chloride in Industry

Sodium chloride (table salt) is used as a seasoning and preservative for foods; in making pottery, soap, glass, textile dyes; and in the production of chemical substances.

If a direct current is passed through a concentrated solution of sodium chloride three main products are obtained (see page 13 for details).

- **Chlorine**, collected at the positive electrode, is used for making chemical substances, bleaches, disinfectants, paints and plastics.
- **Hydrogen**, collected at the negative electrode, is used for manufacturing margarine and ammonia.
- **Sodium hydroxide**, which remains in the solution, is used in making soap, paper and synthetic fibres.

Sodium metal is obtained by electrolysis of molten sodium chloride. This process also produces chlorine (see above for uses of chlorine).

The melting point of the sodium chloride is over 800°C. If calcium chloride is added, the melting point is lowered to about 600°C.

Sodium ions are reduced at the cathode and chloride ions are oxidised at the anode.

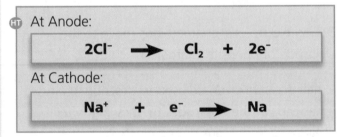

At Anode:

$$2Cl^- \longrightarrow Cl_2 + 2e^-$$

At Cathode:

$$Na^+ + e^- \longrightarrow Na$$

This process of obtaining pure sodium metal from molten sodium chloride uses the Down's cell. A fine iron gauze in the cell stops the reactive chlorine from recombining with the sodium.

A use of sodium metal is in the reduction of titanium(IV) chloride to titanium.

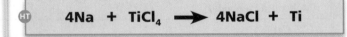

$$4Na + TiCl_4 \longrightarrow 4NaCl + Ti$$

Sodium is used in street lamps and in a certain type of nuclear reactor, transferring heat from the reactor to the steam generators.

Active Electrodes

Sometimes the electrodes take part in the electrolysis process. When this happens then the electrode is called an **active electrode**. This process is used in the industrial purification of copper, in which very pure copper is produced at the cathode from the impure copper anode. This can be seen when copper sulfate solution is electrolysed using copper electrodes.

The masses of the two copper electrodes are measured before they are placed in the electrolyte of copper sulfate. After about 10 minutes the cathode can be seen to grow bigger as it gains copper. The anode can be seen to grow smaller as it loses copper. This is seen clearly when the two electrodes are carefully washed and their masses are re-measured.

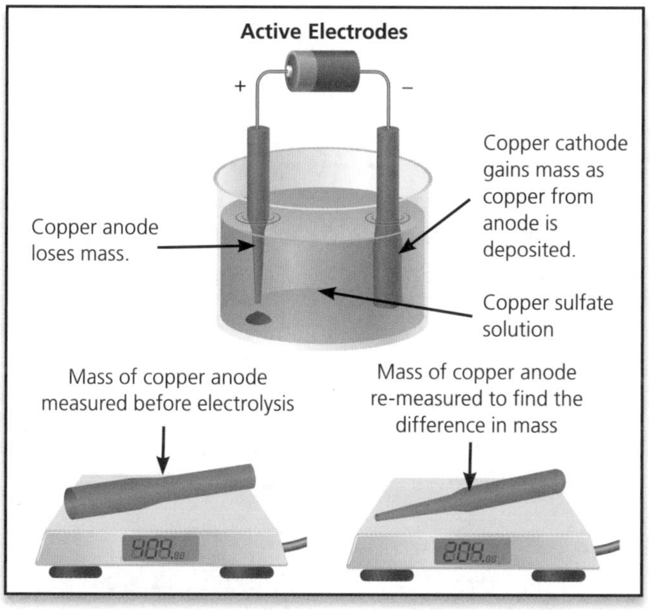
Active Electrodes
+
–
Copper cathode gains mass as copper from anode is deposited.
Copper anode loses mass.
Copper sulfate solution
Mass of copper anode measured before electrolysis
Mass of copper anode re-measured to find the difference in mass

HT At Anode:

$$Cu \rightarrow Cu^{2+} + 2e^-$$

At Cathode:

$$Cu^{2+} + 2e^- \rightarrow Cu$$

When an electrolysis process uses active electrodes:
Loss in mass at anode = Gain in mass at cathode.
This equation is only true if there are no impurities present.

Electroplating

Electroplating is the process by which an object is coated with a thin layer of metal by electrolysis.

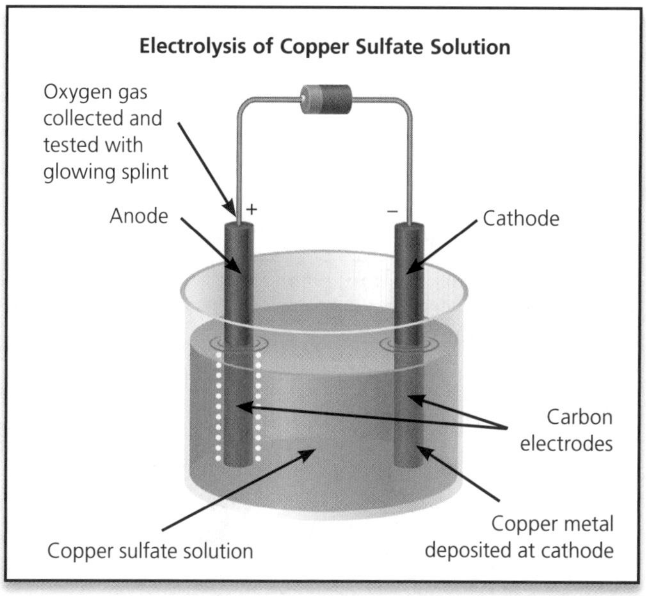
Electrolysis of Copper Sulfate Solution
Oxygen gas collected and tested with glowing splint
Anode
+
–
Cathode
Carbon electrodes
Copper sulfate solution
Copper metal deposited at cathode

HT At Anode:

$$4OH^- - 4e^- \rightarrow 2H_2O + O_2$$

The equation can also be written:

$$4OH^- \rightarrow 2H_2O + O_2 + 4e^-$$

At Cathode:

$$Cu^{2+} + 2e^- \rightarrow Cu$$

Electroplating is quite often used to make an object look attractive and shiny. For example, cheaper metals such as stainless steel (used for cutlery) are sometimes silver plated, and base metals used for jewellery are gold plated.

Electroplating is also used to protect metals against corrosion. Objects made from steel are susceptible to corrosion. For example, steel wheels will rust but if they are chromium plated they look more attractive and they are also protected from rusting (corrosion). Steel food cans will also corrode, which could have health implications, so the cans are coated with a very thin layer of tin by electroplating.

C3 Topic 4: Gases, Equilibria and Ammonia

This topic looks at:

- how to use Avogadro's law and molar volumes of gases
- how equilibrium is used in the Haber process
- why nitrogen-based fertilisers are used
- what happens when nitrogen fertilisers are over-used

HT

Molar Volumes of Gases

The volume of gas produced by a chemical reaction can be measured using a gas syringe.

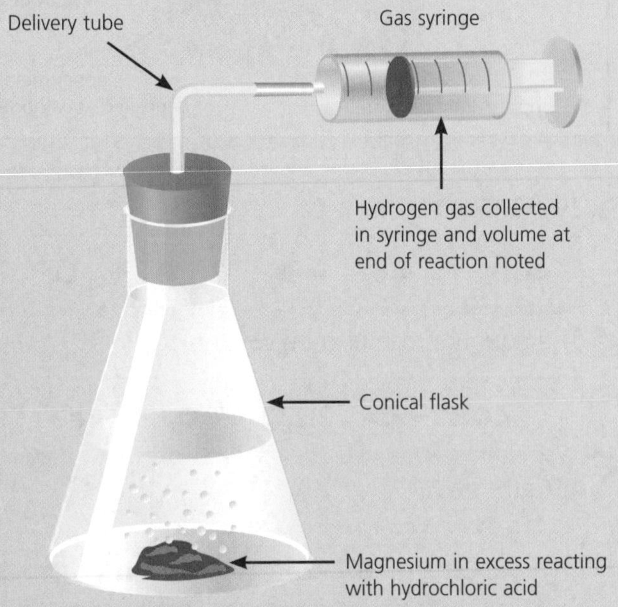
Delivery tube
Gas syringe
Hydrogen gas collected in syringe and volume at end of reaction noted
Conical flask
Magnesium in excess reacting with hydrochloric acid

During the reaction of $100cm^3$ of $1mol\ dm^{-3}$ hydrochloric acid with excess magnesium, $1.2dm^3$ of hydrogen gas was collected in the gas syringe.

Using this data and the equation of the reaction, we can calculate the volume of one mole of hydrogen.

The equation for the reaction is:

$$Mg_{(s)} + 2HCl_{(aq)} \longrightarrow MgCl_{2(aq)} + H_{2(g)}$$

The hydrogen comes from the hydrochloric acid, so write them as a ratio: 2 HCl : 1 H_2.

Convert volume of acid to dm^3:

$$\frac{100}{1000} = 0.1dm^3.$$

Multiply this by concentration used:

$0.1 \times 1 = 0.1$moles.

From the ratio half of this would be hydrogen gas:

$$\frac{0.1}{2} = 0.05\text{moles.}$$

So, 0.05moles of hydrogen has a volume of $1.2dm^3$. Therefore the volume of 1mole of hydrogen is:

$$\frac{1.2}{0.05} = \mathbf{24dm^3}$$

The volume of one mole of any gas, called the **molar volume**, is $24dm^3$ at room temperature and atmospheric pressure. This comes from **Avogadro's law**, which states that equal volumes of all gases, at the same temperature and pressure, contain the same number of molecules.

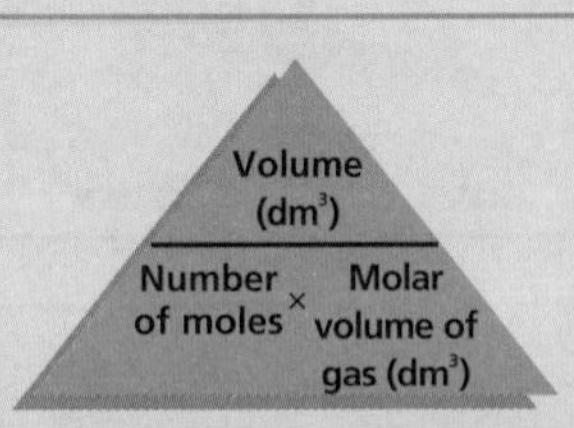
Volume (dm³)
Number of moles
×
Molar volume of gas (dm³)

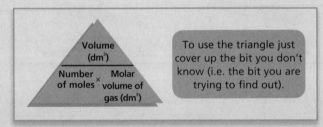

To use the triangle just cover up the bit you don't know (i.e. the bit you are trying to find out).

Example 1

Determine the minimum volume of methane (CH_4) burned completely in oxygen if 11g of carbon dioxide is produced.

(Relative atomic masses: C = 12; O = 16; H = 1)

The word equation for the reaction is:

Methane + Oxygen ⟶ Carbon dioxide + Water

HT

Molar Volumes of Gases (cont.)

The balanced symbol equation for the reaction is:

$$CH_4(g) + 2O_2(g) \rightarrow CO_2(g) + 2H_2O(l)$$

Molar masses:

$$12 + (4 \times 1) + 2 \times (2 \times 16) \rightarrow 12 + (2 \times 16) + 2\,[(2 \times 1) + 16]$$

Reactants → Products:

$$16 + 64 \rightarrow 44 + 36$$

For this example we only need to look at CH_4 and CO_2. So, 16g of CH_4 produces 44g of CO_2.

11g of CO_2 is produced $= \frac{11}{44} = 0.25$mol CO_2

1mol of CH_4 produces 1mol of CO_2, so 0.25mol of CH_4 produces 0.25mol of CO_2 (ratio is 1 : 1).

So, volume of CH_4

= number of moles × molar volume

= 0.25 × 24

= **6dm³**

Example 2

What volume of hydrogen will react with 24dm³ of oxygen to form water?

The equation for the reaction is:

$$2H_2(g) + O_2(g) \rightarrow 2H_2O(l)$$

From the equation, 2 volumes of hydrogen will react with 1 volume of oxygen, or 2 × 24dm³ of hydrogen reacts with 24dm³ of oxygen.

So, **48dm³** of hydrogen will react.

Reversible Reactions

Some chemical reactions are **reversible**. This means that the products will react to give the reactants that we started with.

$$A + B \rightleftharpoons C + D$$

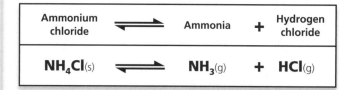

A and B react to produce C and D, but conversely C and D can react to produce A and B.

Ammonium chloride	⇌	Ammonia	+	Hydrogen chloride
$NH_4Cl(s)$	⇌	$NH_3(g)$	+	$HCl(g)$

Solid ammonium chloride decomposes when heated to produce ammonia and hydrogen chloride gas, both of which are colourless.

Reversible Reactions of Ammonium Chloride

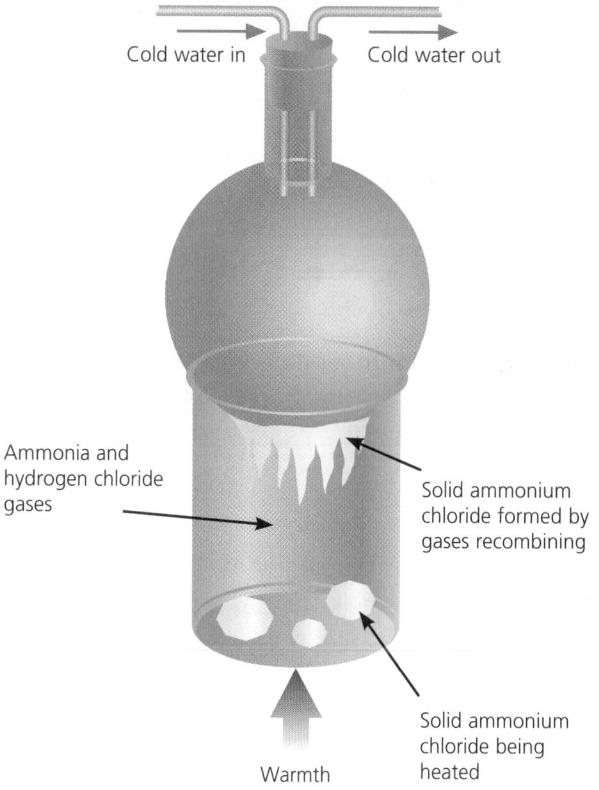
Cold water in
Cold water out
Ammonia and hydrogen chloride gases
Solid ammonium chloride formed by gases recombining
Solid ammonium chloride being heated
Warmth

Ammonia reacts with hydrogen chloride gas to produce clouds of white ammonium chloride powder.

HT Reactions in Equilibrium

Equilibrium is achieved when a reversible reaction occurs in a closed system (where no reactants are added and no products are removed). The reactions occur at exactly the same rate in both directions and never reach completion.

The equilibrium in a reversible reaction will shift to try to counteract any change that is introduced.

Changing Concentration

$$A + B \rightleftharpoons C + D$$

If the concentration of reactant B is increased, the equilibrium will move to counteract this change and reduce the concentration of B again. More reactant A will react with B, producing more C and D.

If the concentration of reactant B is reduced, the equilibrium will move to increase the concentration of B again by reacting more C and D to replace the B that was removed.

Changing Temperature

$$A + B \rightleftharpoons C + D + \text{Heat}$$

Increasing the temperature will cause the position of the equilibrium to move to reduce the temperature again. For example, in a reaction where the forward reaction is exothermic and the back reaction is endothermic, the back reaction will increase to absorb the extra heat and restore the equilibrium.

In the same reversible reaction, if the temperature is decreased, the exothermic forward reaction will increase to generate more heat and restore the equilibrium.

An increase in temperature will always favour the endothermic reaction and a decrease in temperature will always favour the exothermic reaction.

Changing Pressure

$$A_{(g)} + B_{(g)} \rightleftharpoons C_{(g)}$$

Changing pressure will only work in equilibrium reactions involving gases, where the numbers of molecules of reactants and products are different.

If pressure is increased, the equilibrium will move to the side where there are fewer molecules, releasing the pressure and restoring equilibrium.

If pressure is decreased, the reaction will generate more pressure by shifting towards the side with more molecules until equilibrium is restored.

Ammonia

Ammonia is an alkaline gas that is lighter than air and has an unpleasant smell. This gas is produced as part of an intermediate step in the production of ammonium nitrate fertiliser. A fertiliser is either an organic (animal and plant waste) or manmade substance used to make soil more fertile (see page 71).

Making Ammonia

Ammonia has the formula NH_3 so it consists of two elements – nitrogen and hydrogen.

The reaction between nitrogen and hydrogen is reversible. So, as well as nitrogen and hydrogen combining with each other, the ammonia formed will decompose.

Ammonia is made using the Haber process. The raw materials are:

- nitrogen – from the fractional distillation of liquid air
- hydrogen – from natural gas and steam.

HT Making Ammonia (cont.)

If the chemicals are in a closed system, they will reach a stage of **dynamic equilibrium**, i.e. the rate of the forward reaction is equal to the rate of the backward reaction. The word 'dynamic' is used to describe the equilibrium because there is a constant interchange between the reactants and products.

Nitrogen	+	Hydrogen	$\rightleftharpoons$	Ammonia	+	Heat
$N_2(g)$	+	$3H_2(g)$	$\rightleftharpoons$	$2NH_3(g)$	+	Heat

It is important to get the maximum yield in the shortest possible time.

- A **low temperature** increases the yield of ammonia but the reaction is too slow.
- A **high pressure** increases the yield of ammonia but the reaction is too expensive.
- A **catalyst** increases the rate at which equilibrium is reached but does not affect the yield of ammonia.

Fertilisers

The majority of farmers use manmade fertilisers to replace the nitrogen in the soil that has been used up by previous crops. This means that crop yields can be increased. However, manmade fertilisers have advantages and disadvantages.

Advantages

- They are easier to store, distribute and handle than **organic** alternatives.
- They are produced to match crop requirements.
- They help increase crop yield, providing efficiently produced food that costs less.

Disadvantages

- If it gets into drinking water, high nitrate content can be harmful.
- Nitrates can leak into lakes and rivers causing eutrophication.

N.B. These disadvantages are a result of excessive use of manufactured fertilisers.

HT The Haber Process

Fritz Haber showed that ammonia could be made on a large scale, through a process that became known as the Haber process. The conditions chosen are:

- pressure of 200 atmospheres – high pressure 'pushes' the equilibrium position to the right
- temperature of 450°C – in order to give a good rate of ammonia production without making it decompose too much
- iron catalyst – increases the rate at which equilibrium is reached.

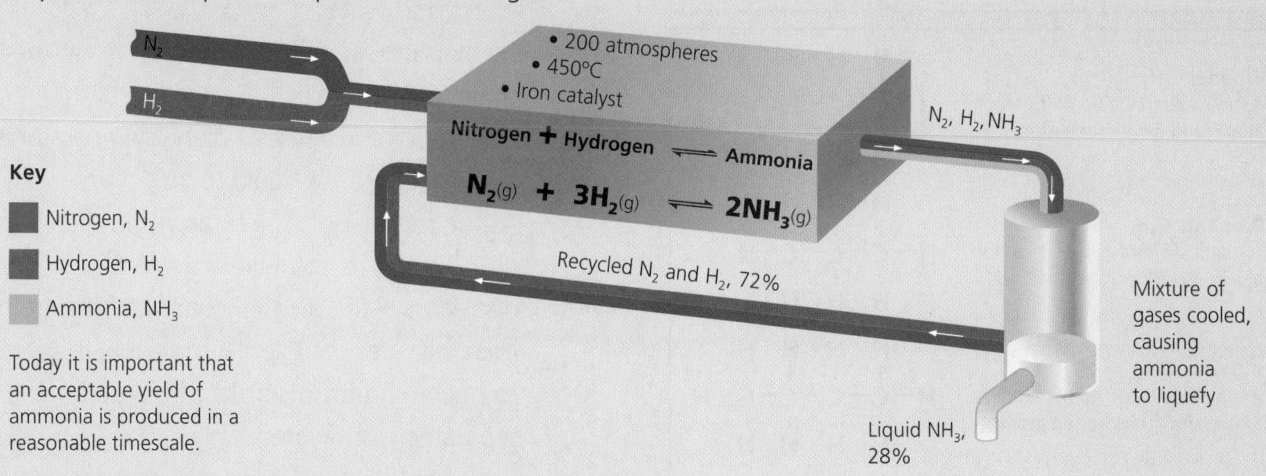
N₂
H₂
• 200 atmospheres
• 450°C
• Iron catalyst
Nitrogen + Hydrogen ⇌ Ammonia
N₂(g) + 3H₂(g) ⇌ 2NH₃(g)
N₂, H₂, NH₃
Recycled N₂ and H₂, 72%
Mixture of gases cooled, causing ammonia to liquefy
Liquid NH₃, 28%
Key
Nitrogen, N₂
Hydrogen, H₂
Ammonia, NH₃

Today it is important that an acceptable yield of ammonia is produced in a reasonable timescale.

C3 Topic 5: Organic Chemistry

This topic looks at:

- the properties of alkanes and alkenes
- the properties of alcohols and carboxylic acids
- the social and health effects of alcohol
- how fermentation can produce ethanol
- the methods for producing concentrated and pure ethanol
- the reactions and uses of ethanoic acid and esters
- what happens during the hydrogenation of oils
- how soap works

Homologous Series

A homologous series is a series of chemicals with a similar general formula and similar chemical properties. For example, alkanes all have the formula C_nH_{2n+2} and alkenes all have the formula C_nH_{2n}. The physical properties of members of these homologous series gradually change as the length of the carbon chain increases. For example, the alkane **methane** is a gas with a low boiling point, while **octane** is a liquid with a higher boiling point.

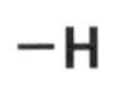

Hydrogen atoms can make 1 bond each.

Carbon atoms can make 4 bonds each.

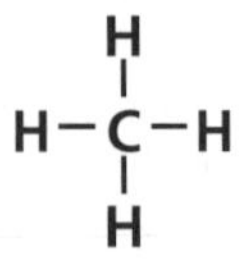

The simplest alkane, **methane, CH_4,** is made up of 4 hydrogen atoms and 1 carbon atom.

Ethane, C_2H_6
A molecule made up of 2 carbon atoms and 6 hydrogen atoms.

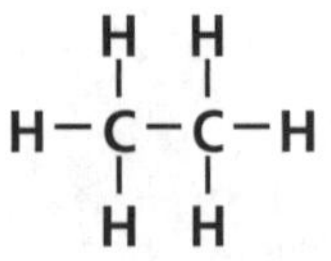

Propane, C_3H_8
A molecule made up of 3 carbon atoms and 8 hydrogen atoms.

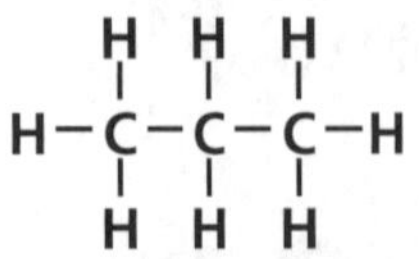

Butane, C_4H_{10}
A molecule made up of 4 carbon atoms and 10 hydrogen atoms.

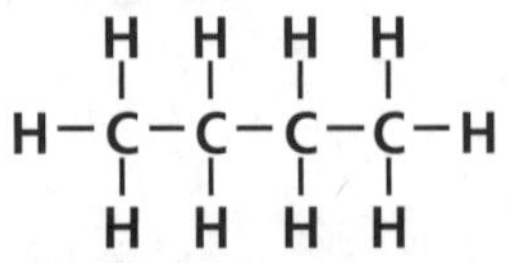

The formulae and structures of the first two alkenes are shown below.

Ethene, C_2H_4
A molecule made up of 4 hydrogen atoms and 2 carbon atoms. Ethene contains 1 double carbon–carbon bond.

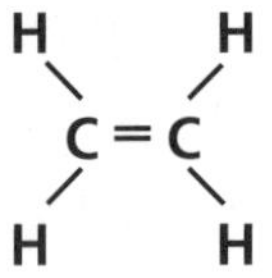

Propene, C_3H_6
A molecule made up of 6 hydrogen atoms and 3 carbon atoms. Propene contains 1 double carbon–carbon bond and 1 single carbon–carbon bond.

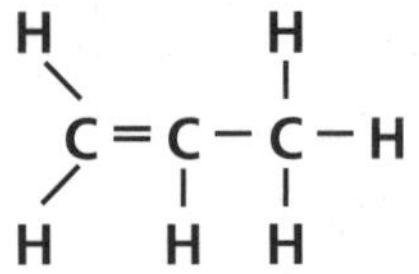

HT Alcohols

Alcohols are organic compounds. The simplest **alcohols** are as follows:

Methanol
CH_3OH

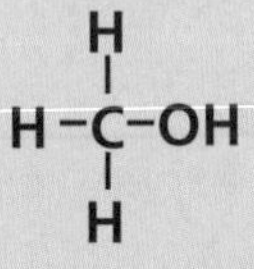

Ethanol
C_2H_5OH

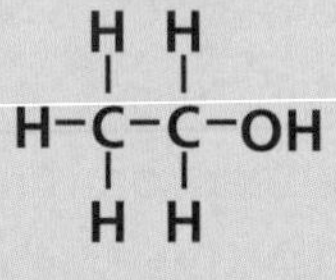

Propanol
C_3H_7OH

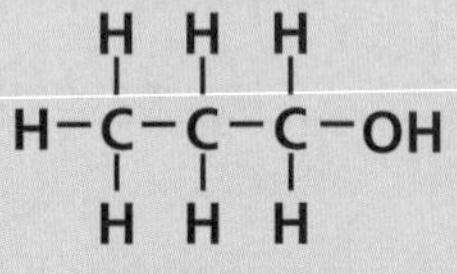

Alcohols all contain a particular group of atoms: –OH (hydroxyl). This is the functional group, and it is this that gives alcohols their characteristic properties.

Properties of Alcohols

- Alcohols that have small molecules, such as methanol, ethanol and propanol, have characteristic odours.
- They react with oxygen to form carboxylic acids.
- They react with carboxylic acids to form esters and water.
- As the number of the carbon atoms in the chain increases (i.e. 1, 2, 3, 4, etc.), the boiling point increases – higher temperatures are needed for the liquid to boil.
- As the number of carbon atoms in the chain gets bigger, its solubility in water decreases – the alcohol changes from being **miscible** (i.e. completely mixed) with water to becoming **immiscible** with water (i.e. separated from water).

Uses of Alcohols

Some of the many uses for alcohols include:

- making **esters** – sweet-smelling compounds used to make fragrances and food flavourings (see page 78)
- making **emulsifiers**, **emollients** (softening agents) and **thickeners** for cosmetics
- making **solvents** – the simplest alcohols (methanol, ethanol and propanol) are used as solvents. For example, ethanol is used as a solvent for fragrances in perfumes, and propan-2-ol is used as a solvent for dyes and pigments in ink-jet printer ink, cosmetics and for food flavourings.

MASCARA

Ethanol

Ethanol is the form of alcohol that is found in alcoholic drinks. It can have many harmful effects on the body. Prolonged consumption of alcoholic drinks (ethanol) could result in these effects being permanent. Effects include:

- a deficiency in vitamin B, causing skin damage, diarrhoea and depression
- decreased levels of iron, leading to anaemia
- liver damage – the liver would no longer be able to carry out its vital functions of making toxins safe
- destruction of brain cells
- an increased risk of cancer of the mouth, larynx, oesophagus, liver, stomach, colon, rectum and possibly the breast
- an increased risk of heart disease and high blood pressure
- inflammation and irritation of the intestinal and stomach lining, leading to ulcers and damage to the pancreas
- in men, an inability to get an erection, shrinking testes and penis and a reduced sperm count
- in women, disruption to the menstrual cycle, risk of miscarriage and low birth weight or birth defects in their babies.

Alcohol and Society

Binge drinking occurs when a lot of alcohol is consumed in one session. There has been an increase in cases of binge drinking in the UK over recent years. This has many social implications, for example:

- hangovers
- mood changes that can make people tearful or aggressive
- people finding themselves in violent situations (where perhaps they would not if they had not been drinking)
- increased risk of accidents – wasting time, space in hospitals, doctors' surgeries
- lowered inhibitions leading to greater risks of getting into trouble or doing things the person may regret afterwards.

Ethanol by Fermentation

Any substance that contains sugar can be fermented. To do this, the sugar (glucose) is mixed with yeast in the absence of air.

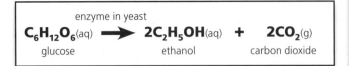

$$\underset{\text{glucose}}{C_6H_{12}O_6(aq)} \xrightarrow{\text{enzyme in yeast}} \underset{\text{ethanol}}{2C_2H_5OH(aq)} + \underset{\text{carbon dioxide}}{2CO_2(g)}$$

An enzyme in the yeast breaks down the glucose into ethanol and carbon dioxide.

Concentration of Alcohol

Ethanol that is made by fermentation results in a low concentration of ethanol in solution. This is because when the concentration of ethanol reaches levels of 16% to 18% by volume of solution, the enzymes in yeast become **denatured**.

Alcohols that are available for human consumption have different concentrations. Most beers contain about 3% to 6% of ethanol. Wines usually contain a concentration of 12.5% to 14.5% ethanol and fortified wines such as sherry contain a concentration of 18% to 22% ethanol.

If the ferment is then distilled (see page 75), beverages with a higher concentration of ethanol can be produced. For example, some spirits have a concentration of 40% to 50% ethanol.

HT Making Ethanol from Ethene

Ethene can be reacted with water in the presence of a catalyst (phosphoric acid) to produce ethanol.

$$\text{Ethene} + \text{Water (as steam)} \xrightleftharpoons{\text{Phosphoric acid}} \text{Ethanol}$$

$$C_2H_4(g) + H_2O(l) \xrightleftharpoons{\text{Phosphoric acid}} C_2H_5OH(g)$$

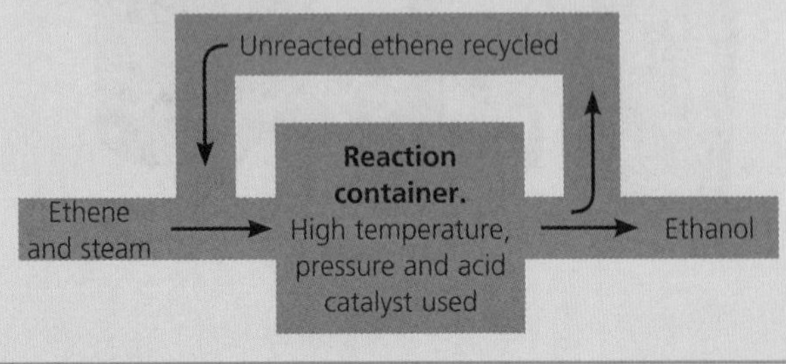
Unreacted ethene recycled
Ethene and steam
Reaction container.
High temperature, pressure and acid catalyst used
Ethanol

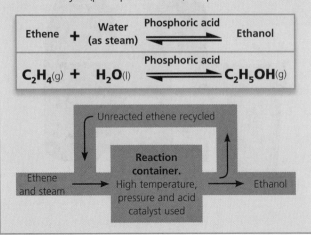

HT Comparing Methods

There are two main methods of producing ethanol. One method, fermentation, produces impure ethanol but uses renewable resources. The other produces pure ethanol but uses non-renewable resources.

The following table shows the advantages and disadvantages for both methods of production.

Fermentation of Carbohydrates

Advantages
- Renewable as raw material is sugar cane, maize, corn, rice, etc.
- Cheap labour is readily available.
- Very little energy is used during production (maintain temperature at 30–40°C).

Disadvantages
- Large areas of land are needed to grow crops.
- Carbon dioxide is produced.
- Batch process, so is very labour intensive.
- Slow reaction.
- Low concentration of ethanol is produced. It must be purified by fractional distillation.

Reacting Ethene with Steam

Advantages
- Continuously produces ethanol.
- Fewer workers needed.
- Fast process.
- Pure ethanol is produced.

Disadvantages
- Uses non-renewable crude oil as raw material.
- Uses high amounts of energy to maintain the temperature at 300°C.

Separating Ethanol by Fractional Distillation

Ethanol can be made by fermentation of sugars, but this only produces a mixture of water and ethanol with a low concentration of ethanol. A more concentrated mixture of ethanol can be obtained by using fractional distillation.

Fractional Distillation of Ethanol

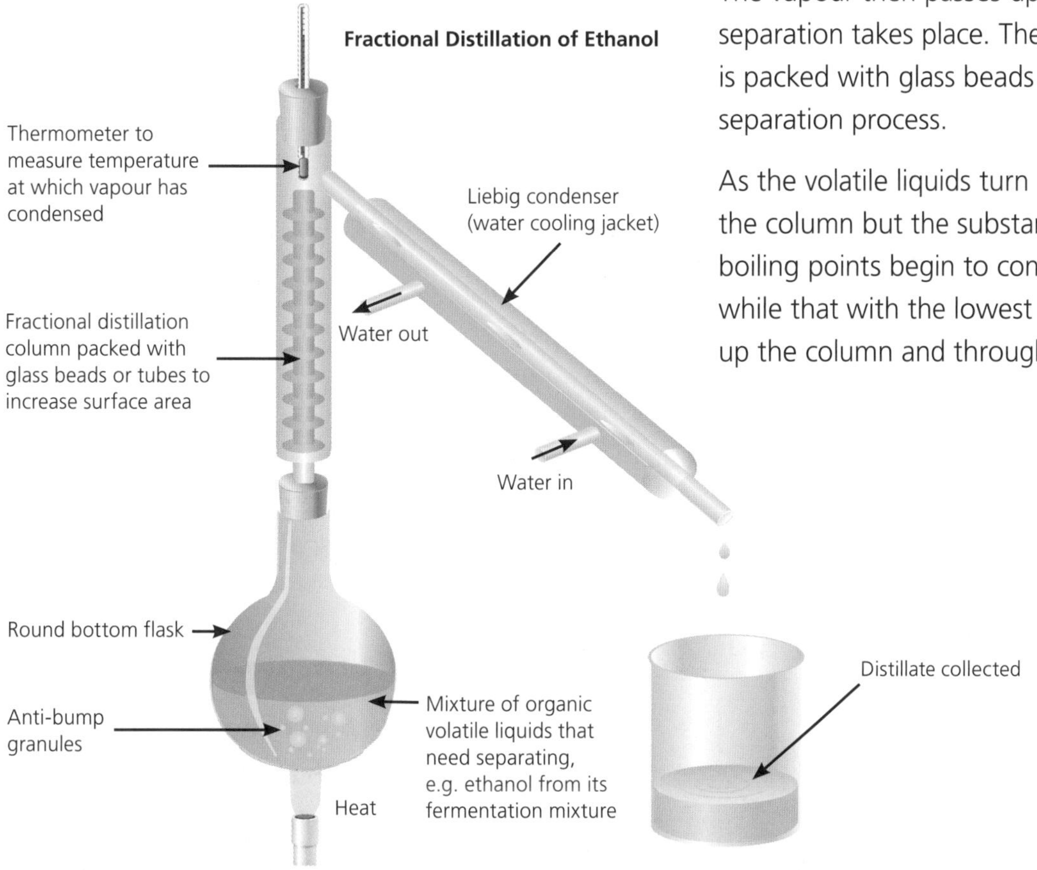
Thermometer to measure temperature at which vapour has condensed
Liebig condenser (water cooling jacket)
Water out
Fractional distillation column packed with glass beads or tubes to increase surface area
Water in
Round bottom flask
Distillate collected
Anti-bump granules
Mixture of organic volatile liquids that need separating, e.g. ethanol from its fermentation mixture
Heat

The liquid mixture containing the ethanol is boiled. The liquids all have different boiling points so the most volatile will boil first.

The vapour then passes up the column where the separation takes place. The fractionating column is packed with glass beads to help the separation process.

As the volatile liquids turn into vapour they rise up the column but the substances with the higher boiling points begin to condense on the glass beads, while that with the lowest boiling point rises further up the column and through the condenser.

HT Dehydration of Ethanol

Ethanol can be dehydrated (have the water removed) to make ethene. Dehydrating ethanol involves passing ethanol vapour over a heated catalyst and collecting the resultant ethene over water.

Ethanol	$\xrightarrow[\text{Catalyst}]{\text{Heat}}$	**Ethene**	**+**	**Water**
$C_2H_5OH_{(g)}$	$\xrightarrow[Al_2O_3]{}$	$C_2H_{4(g)}$	**+**	$H_2O_{(l)}$
H–C(H)(H)–C(H)(H)–O–H	$\xrightarrow{\text{Catalyst}}$	H₂C=CH₂ (structural)	**+**	H–O–H

Dehydration of Ethanol

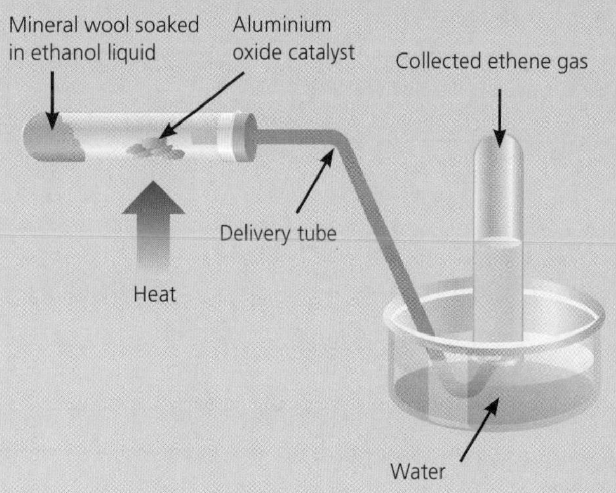
Mineral wool soaked in ethanol liquid
Aluminium oxide catalyst
Collected ethene gas
Delivery tube
Heat
Water

Oxidation of Ethanol

When ethanol is oxidised it produces vinegar in solution, for example, red wine vinegar. (Red wine vinegar is made when red wine oxidises with oxygen in the air.) This means that ethanol has reacted with oxygen and produced an **organic acid** – in this instance ethanoic acid and water.

The equation is:

Ethanol + Oxygen → Ethanoic acid + Water

$$C_2H_5OH_{(l)} + O_{2(g)} \rightarrow CH_3COOH_{(aq)} + H_2O_{(l)}$$

HT

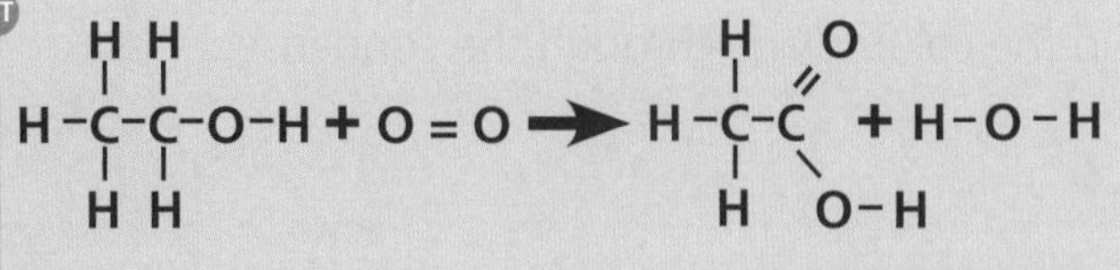

HT

Carboxylic Acids

The most common organic acids are **carboxylic acids**, which all have the functional group –COOH (carboxyl). The three simplest carboxylic acids are as shown below:

Methanoic acid 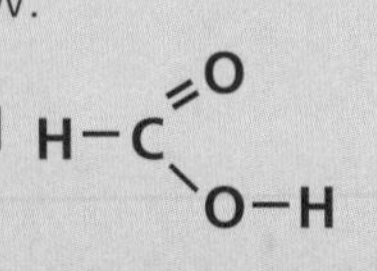 HCOOH

Ethanoic acid 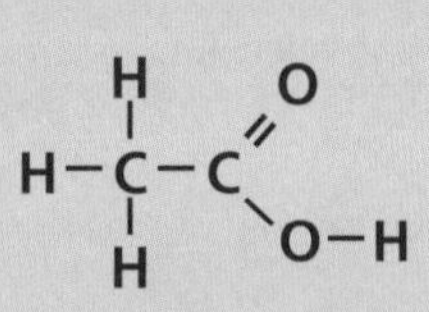 CH_3COOH

Propanoic acid 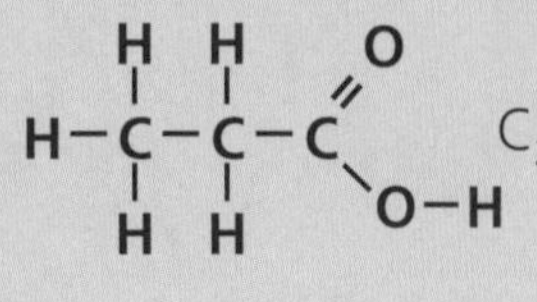 C_2H_5COOH

Properties of Organic Acids

- Organic acids will react with alcohols to form **esters**.
- As the number of carbon atoms in the chain increases:
  – the acid changes from a colourless liquid to a **waxy solid**
  – the **odour** becomes less pungent
  – **solubility** in water decreases, so, like alcohols, the acid becomes more **immiscible** in water.

Uses of Organic Acids

Organic acids have many uses. For example, they can be used:

- as solvents
- in medicines
- in some food items (e.g. vinegar, which is used as a flavouring and a preservative)
- in the manufacture of soaps and detergents (see page 79).

Reactions of Ethanoic Acid

Like all organic acids, ethanoic acid will react with alcohols to form an ester and water. For example:

Ethanoic acid + Ethanol ⇌ Ethyl ethanoate + Water

HT $$CH_3COOH_{(l)} + C_2H_5OH_{(l)} \rightleftharpoons CH_3COOC_2H_{5(l)} + H_2O_{(l)}$$

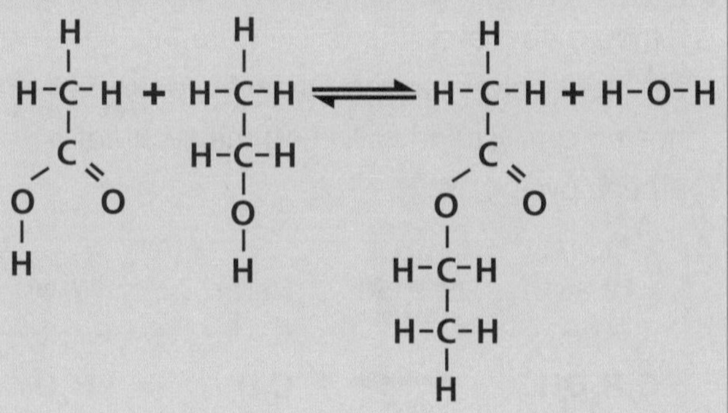

Reactions of Ethanoic Acid (cont.)

Reaction with an alkali, e.g. KOH

Acid	+	Alkali	→	Salt	+	Water
Ethanoic acid	+	Potassium hydroxide	→	Potassium ethanoate	+	Water
CH_3COOH(aq)	+	KOH(aq)	→	$CH_3COO^-K^+$(aq)	+	H_2O(l)

HT

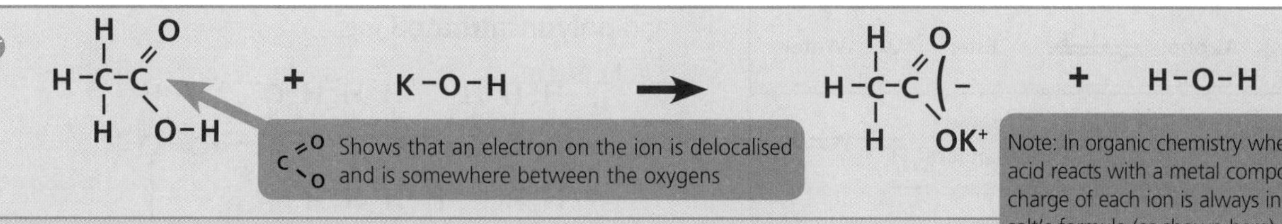
Shows that an electron on the ion is delocalised and is somewhere between the oxygens

Note: In organic chemistry when an organic acid reacts with a metal compound, the charge of each ion is always included in the salt's formula (as shown here).

Reaction with an oxide, e.g. CaO

Acid	+	Oxide	→	Salt	+	Water
Ethanoic acid	+	Calcium oxide	→	Calcium ethanoate	+	Water
$2CH_3COOH$(aq)	+	$Ca^{2+}O^{2-}$(s)	→	$(CH_3COO^-)_2Ca^{2+}$(aq)	+	H_2O(l)

HT

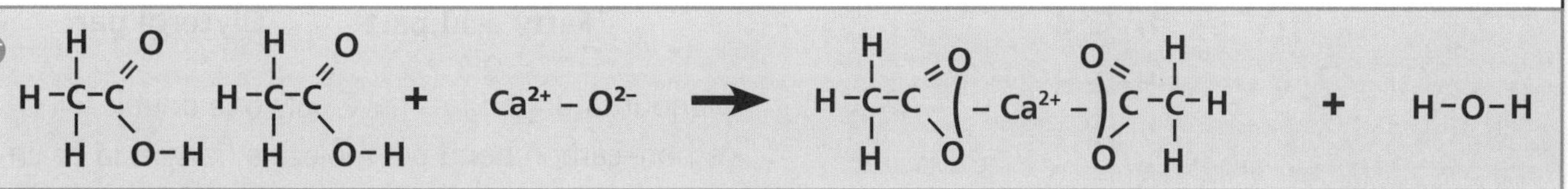

Reaction with metals, e.g. Mg

Acid	+	Metal	→	Salt	+	Hydrogen
Ethanoic acid	+	Magnesium	→	Magnesium ethanoate	+	Hydrogen
$2CH_3COOH$(aq)	+	Mg(s)	→	$(CH_3COO^-)_2Mg^{2+}$(aq)	+	H_2(g)

HT

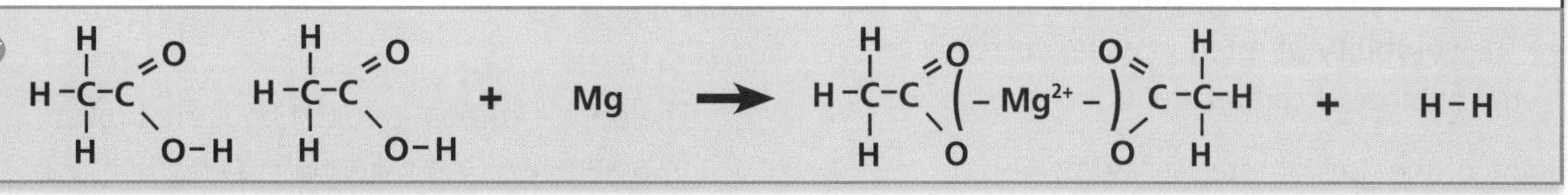

Reaction with a carbonate, e.g. Na_2CO_3

Acid	+	Carbonate	→	Salt	+	Carbon dioxide	+	Water
Ethanoic acid	+	Sodium carbonate	→	Sodium ethanoate	+	Carbon dioxide	+	Water
$2CH_3COOH$(aq)	+	Na_2CO_3(s)	→	$2CH_3COO^-Na^+$(aq)	+	CO_2(g)	+	H_2O(l)

HT

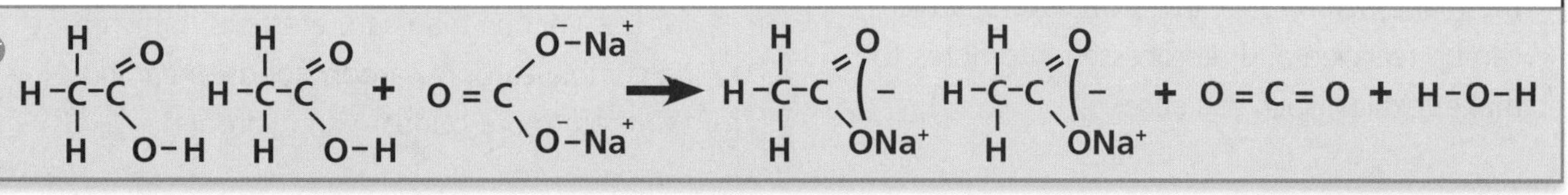

Esters

Esters are organic compounds that contain carbon, hydrogen and oxygen. They are produced from the reaction between an **organic acid** and an **alcohol**. The reaction produces an ester and water.

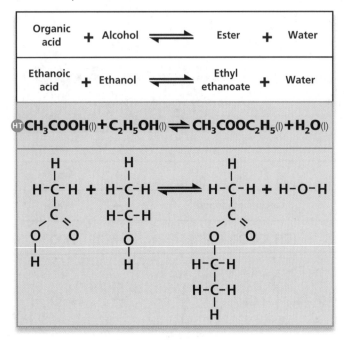

For example:

Organic acid	+	Alcohol	⇌	Ester	+	Water
Ethanoic acid	+	Ethanol	⇌	Ethyl ethanoate	+	Water

HT $CH_3COOH_{(l)} + C_2H_5OH_{(l)} \rightleftharpoons CH_3COOC_2H_{5(l)} + H_2O_{(l)}$

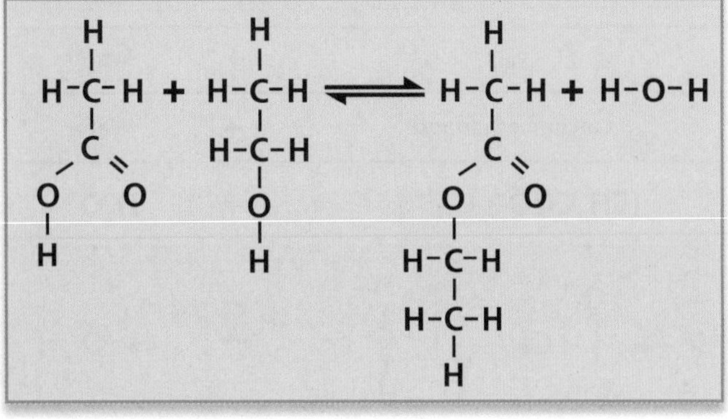

Properties and Uses of Esters

- Esters are **soluble** in organic solvents (and are good solvents themselves).
- They vary from colourless volatile **liquids** to colourless waxy **solids**.
- All esters have **strong fruity smells**.
- The **solubility** of esters in water decreases as the number of carbon atoms increases.

Some of the uses of esters include:

- as **fruit flavourings**, for example, methyl methanoate is found in raspberry essence
- in **cosmetics** – fatty acid esters are used in cosmetics, for example, as **emollients** (substances that soothe and soften the skin) and as perfumes (as they are sweet smelling).
- the manufacture of simple **polyesters**, which can be recycled and reprocessed into fibres to make textiles, such as fleece).

Oils

Animal fats and plant oils are naturally occurring organic compounds and are based on esters made from an alcohol called glycerol. A typical ester found in fats and oils is glyceryl stearate. Vegetable oils are **unsaturated**. They are liquids at room temperature because they have high levels of **monounsaturated** and **polyunsaturated** fats.

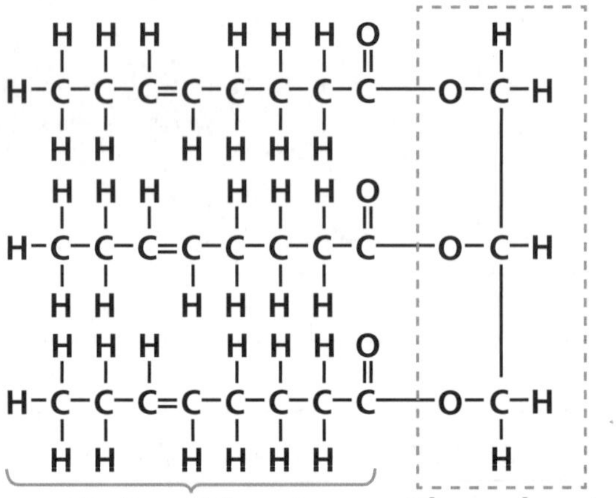
Fatty acid part
Glycerol part

Monounsaturated fats have only one double carbon–carbon bond per molecule (fatty acid chain).

Polyunsaturated fats have two or more double carbon–carbon bonds per molecule (fatty acid chain). Removing some or all of the double carbon–carbon bonds will raise the melting point of an oil and, therefore, harden it. This can have the advantage of increasing its shelf life, which is an advantage when making margarine.

HT The double carbon–carbon bonds are broken by adding hydrogen. Enough double bonds have to be **hydrogenated** to achieve the right viscosity (thickness). This process involves catalytic hydrogenation and is used in the production of margarines. The more hydrogen that is added, the greater the viscosity.

Monounsaturated and polyunsaturated oils are far less viscous than saturated oils, which have no double carbon–carbon bonds, meaning there are more bonds with hydrogen.

Making Soaps

Soap was the first **detergent** discovered. It is used to help dissolve oil in water.

Soaps are made from plant oils or animal fats. Plant oils and animal fats contain **fatty acids**.

These oils and fats are converted into soap by boiling them with the alkali compounds, potassium hydroxide or sodium hydroxide solution. The reaction produces glycerol from the fatty acids, and salts of the fatty acids. This process is called **saponification**.

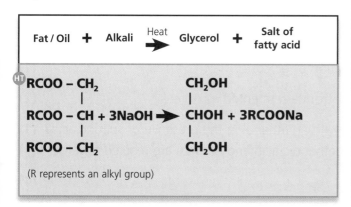

Fat / Oil + Alkali $\xrightarrow{\text{Heat}}$ **Glycerol + Salt of fatty acid**

HT

$$\begin{array}{lll} RCOO-CH_2 & & CH_2OH \\ \quad\;\;| & & \;\;| \\ RCOO-CH + 3NaOH & \rightarrow & CHOH + 3RCOONa \\ \quad\;\;| & & \;\;| \\ RCOO-CH_2 & & CH_2OH \end{array}$$

(R represents an alkyl group)

HT

Detergents

Detergents work because they act like **emulsifiers**. Each detergent has a **hydrophilic** (water-loving) end and a **hydrophobic** (water-hating) end.

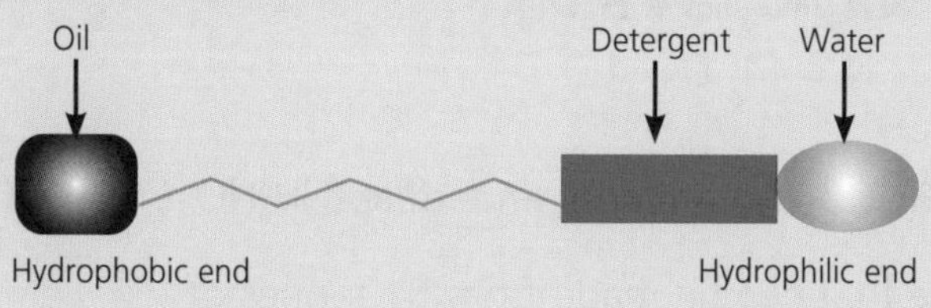
Oil
Detergent
Water
Hydrophobic end
Hydrophilic end

Oils will not mix with water, i.e. the oil is **immiscible** in water. This means that oil stains do not easily come off crockery and cutlery, or out of fabric.

Detergents help the oils to mix with water, making an emulsion that is **miscible** with water. This makes oils easier to clean off.

HT

Surface Tension

Surface tension can be described as the surface of the liquid behaving like an elastic sheet when it is in contact with air. The intermolecular forces between the liquid particles constantly pull the particles towards each other. The particles that are along the surface are always being pulled towards the rest of the liquid by these forces, so the liquid's surface behaves like an elastic sheet.

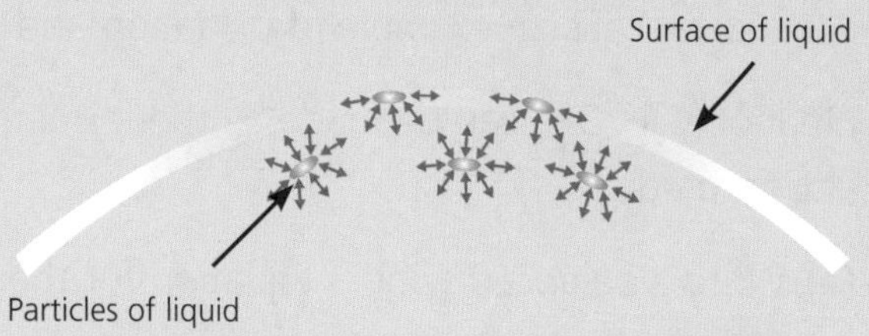
Surface of liquid
Particles of liquid

All types of detergent contain substances called **surfactants**. Surfactants lower the surface tension of water, which means that oil can be pulled away from the crockery, cutlery, fabric, etc.

Water Under Surface Tension

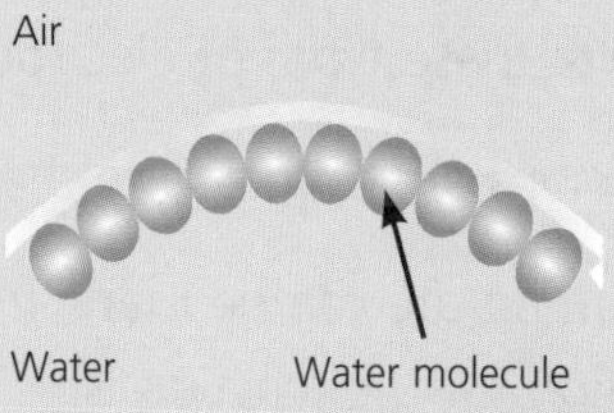
Air
Water
Water molecule

Detergent Lowers Surface Tension

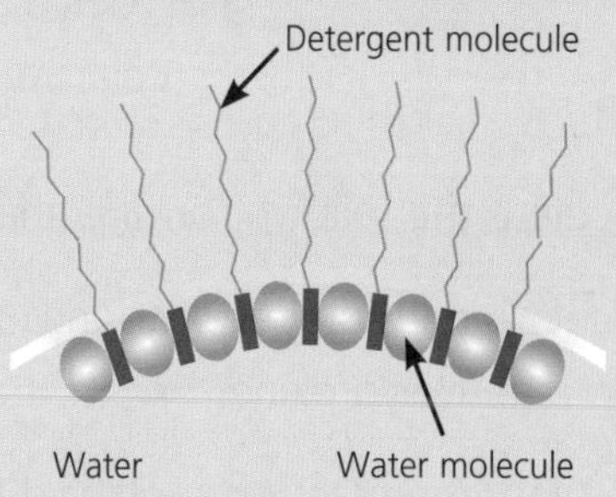
Detergent molecule
Water
Water molecule

When a detergent cleans an item:

- the water-hating end sticks in the oil or grease
- the water-loving end is still in the water
- the oil or grease is pulled off the item into the water.

Exam Practice Questions

Questions labelled with an asterisk (*) are ones where the quality of your written communication will be assessed – you should take particular care with your spelling, punctuation and grammar, as well as the clarity of expression, on these questions.

1 **(a)** Ethanol belongs to a homologous series, called alcohols. Describe what a homologous series is. **(2)**

(b) Write the structural formula of the two simplest alkenes. **(2)**

(c) Give three uses of alcohols. **(3)**

2 **(a)** In electrolysis, what is oxidation? **(1)**

(b) In electrolysis, what is reduction? **(1)**

(c) Explain in terms of transfer of electrons why the electrolysis of an aqueous or molten ionic substance is an example of oxidation and reduction reactions. **(4)**

3 Ethanol (C_2H_5OH) can be made by the fermentation of glucose ($C_6H_{12}O_6$). Carbon dioxide (CO_2) is also formed.

(a) Write a balanced symbol equation for the fermentation of glucose. **(1)**

(b) Describe how you can obtain concentrated ethanol from a fermented mixture. **(2)**

(c) (i) What is made when an organic acid is reacted with an ethanol in the presence of sulfuric acid? **(1)**
(ii) State the name of the product produced when ethanoic acid and ethanol are reacted together. **(1)**

4 **(a)** What process is used to purify copper? **(1)**

(b) What type of electrodes are used during the purification of copper? **(1)**

(c) How are these types of electrodes different from those used in standard electrolysis processes? **(2)**

(d) Which ions are attracted to the cathode and why? **(2)**

5 Ammonium nitrate can be made from the reaction between nitric acid and ammonium hydroxide. The equation for the reaction is:

$$\mathbf{NH_4OH}(aq) + \mathbf{HNO_3}(aq) \longrightarrow \mathbf{NH_4NO_3}(aq) + \mathbf{H_2O}(l)$$

(Relative atomic masses: H = 1; N = 14; O = 16)

(a) Calculate the relative formula masses for nitric acid (HNO_3) and ammonium nitrate (NH_4NO_3). **(2)**

(b) Calculate the number of tonnes of ammonium nitrate that can be made from 51 tonnes of ammonium hydroxide. **(2)**

(c) Calculate the mass of nitric acid that would be needed to make 56 tonnes of ammonium nitrate. **(2)**

(d) Calculate the percentage by mass of nitrogen in ammonium nitrate. **(2)**

6 **(a)** Why is more soap needed in hard water? **(3)**

(b) Which of these substances most resemble detergents?

A ☐ an emulsifier **B** ☐ soap **C** ☐ an oil **D** ☐ a solvent **(1)**

7 **(a)** Describe, with an example, how esters are made. Your answer must include a word equation for the reaction. **(2)**

(b) Give three properties of esters that make them useful. **(3)**

HT

8 **(a)** Under what conditions does a mole of gas occupy a volume of $24dm^3$? **(1)**

(b) For the conditions stated in **8(a)**, calculate the volume of gas in each of the following:

(i) 0.25 moles of sulfur dioxide gas **(1)**

(ii) 3 moles of nitrogen gas **(1)**

(iii) 0.5 moles of oxygen gas **(1)**

(iv) 1.25 moles of methane gas **(1)**

9 *Explain in as much detail as possible why it is important for industry to use a catalyst, a change of temperature and, quite often, a change in reactant concentration (or pressure, if gases) to alter the equilibrium of a reversible reaction. **(6)**

10 **(a)** When a reaction between gases has its pressure increased how is the position of equilibrium affected? **(1)**

(b) If a forward reaction takes in more heat what will the reverse reaction do? **(1)**

(c) What happens if the reaction is carried out in a closed system? **(1)**

(d) Explain what happens to a reversible reaction in equilibrium when the conditions are changed. **(2)**

11 **(a)** A solution of copper chloride ($CuCl_2$) has a concentration of $0.4mol\ dm^{-3}$. How much copper chloride is there in $0.75dm^3$ of solution? **(2)**

(b) What is the mass of the solute, copper chloride, present in the solution? **(2)**
(Relative atomic masses: Cu = 63.5; Cl = 35.5)

(c) $0.25dm^3$ of a solution of copper sulfate contains 0.3 moles of copper sulfate. Calculate its concentration. **(1)**

(d) What is the mass of copper sulfate that is needed to make this solution? **(2)**
(Relative atomic masses: Cu = 63.5; S = 32; O = 16)

12 State, with reasons, which gas, $200cm^3$ of helium or $300cm^3$ of sulfur dioxide, contains the most particles under the same conditions of temperature and pressure. **(3)**

13 The following equation shows what happens in a titration reaction:

$$\mathbf{2NH_4OH_{(aq)} + H_2SO_{4(aq)} \longrightarrow (NH_4)_2\,2SO_{4(aq)} + H_2O_{(l)}}$$

$0.025dm^3$ of $0.8mol\ dm^{-3}$ sulfuric acid is found to neutralise $0.15dm^3$ of ammonium hydroxide.

Calculate the concentration of the ammonium hydroxide. **(3)**

Answers

Model answers have been provided for the quality of written communication questions that are marked with an asterisk (*). The model answers would score the full 6 marks available. If you have made most of the points given in the model answer and communicated your ideas clearly, in a logical sequence with few errors in spelling, punctuation and grammar, you would get 6 marks. You will lose marks if some of the points are missing, if the answer lacks clarity and if there are serious errors in spelling, punctuation and grammar.

Unit C1

1. (a) B
(b) A salt

2. (a) The Earth's early atmosphere was mostly composed of carbon dioxide **(1 mark)**; Over time levels of carbon dioxide decreased and levels of nitrogen and oxygen increased **(1 mark)**
(b) When the Earth and its atmosphere cooled, water vapour released into the atmosphere by volcanic eruptions condensed to become liquid **(1 mark)**; This fell as rain and filled up the hollows in the Earth's crust **(1 mark)**

3. Any three from: Through burning fossil fuels; Deforestation; An increase in cattle farming / rice growing; Motorised transportation

4. (a) A compound made up of hydrogen and carbon only
(b) Any one from: A greater molecular mass; Higher melting/boiling point; More viscous; Less easy to ignite
(c) Because it contains lots of different hydrocarbon molecules of different sizes
(d) B

5. (a) Any one from: Cuts down on excavation / mining; Uses less water and chemicals; Saves on raw materials; Less energy required
***(b)** Recycling more materials such as metals, paper, glass and plastics can be considered to be sustainable because it helps to conserve natural resources, it reduces the amount of energy needed to turn the used material into a new product, and it provides more employment opportunities within the ever-increasing recycling industry. By recycling these materials there will be less future need in economic investment to find new replacement materials or new sources of the material. This also means that the metals, paper, glass and plastics will not be left on a landfill site, causing potential environmental risks and damaging any future use of that area.

6. (a) An ore is a naturally occurring mineral in the Earth's crust that contains compounds of metals.
(b) A metal's position in the reactivity series determines how easily it is extracted.
(c) (i) Copper; Carbon dioxide
(ii) It has been reduced to copper metal.
(d) Loss of oxygen or gain of electrons in a compound during a chemical reaction.

7. Aluminium can only be extracted by electrolysis **(1 mark)**; Aluminium oxide cannot be reduced by carbon because carbon is less reactive than aluminium **(1 mark)**; Iron is extracted by heating the iron oxide with carbon **(1 mark)** because carbon is more reactive than iron **(1 mark)**

8. (a) Calcium oxide; Carbon dioxide
(b) If heated with clay, it will produce cement, which can be used to make concrete or plaster **(1 mark)**; If heated with sand, it will produce glass **(1 mark)**
(c) At least one point from environmental, economic and social, plus any additional point, from:
Environmental: The effect on native animal habitats / landscape; Noise and air pollution; Additional traffic
Economic: Costs involved in quarrying and processing; Effect on local businesses
Social: Availability of workforce; How the quarry can be used afterwards

9. (a) Sulfuric acid + Magnesium oxide ⟶ Magnesium sulfate + Water
(b) Nitric acid + Copper oxide ⟶ Copper nitrate + Water
(c) The stomach contains hydrochloric acid, which reacts with the base **(1 mark)** in the indigestion remedy to form a salt and water **(1 mark)**
(d) To kill harmful bacteria on any food consumed **(1 mark)**; To give enzymes their optimum environment for digestion **(1 mark)**

10. (a) A monomer is a short-chain hydrocarbon **(1 mark)**; A polymer is a large molecule made up of a repeating pattern of identical smaller chemical molecules called monomers **(1 mark)**
(b) 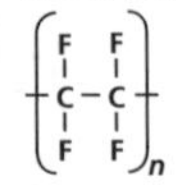
(c) B
(d) Burning polymers produces air pollution **(1 mark)**; The carbon dioxide that is also produced contributes to the greenhouse effect **(1 mark)**; When burned some polymers produce toxic fumes **(1 mark)**
(e) Any one from: Water-resistant; Flexible; Non-stick; Unreactive

11. (a) Ions in the brine are free to move around whilst in solution **(1 mark)**; Electric (d.c.) current passed through the electrolyte causes positive hydrogen ions to flow to the negative electrode forming hydrogen gas **(1 mark)**; and negative chloride ions to flow to the positive electrode forming chlorine gas **(1 mark)**
(b) (i) A greenish-yellow gas
(ii) Test using damp litmus paper, which will be bleached by the chlorine
(c) Gas is poisonous **(1 mark)**; Would cause an environmental hazard if large quantities released into the atmosphere **(1 mark)**

12.* The atoms in pure iron are all the same size and in a regular arrangement. By mixing iron with other elements it becomes an alloy. In an alloy the regular arrangement of the iron atoms (ions) is disrupted because the atoms of the alloying element atoms are either larger or smaller than the iron atoms (ions). This gives alloys improved properties over the pure metal, for example, mixing carbon with iron makes steel. In steel the carbon atoms interrupt the layers of iron atoms (ions) so they are no longer able to move easily over each other, making the alloy much harder and stronger so it can be used in the construction industry.

13. (a) (i) $CH_4(g) + 2O_2(g) \longrightarrow CO_2(g) + 2H_2O(l)$
(ii) **For example:** Methane + Oxygen ⟶ Carbon dioxide + Carbon monoxide + Water **(1 mark)**
$3CH_4 + 5O_2 \longrightarrow CO_2 + 2CO + 6H_2O$ **(1 mark)**
(see additional reactions on page 19)
*(b) Supply refers to the amount of petrol or diesel oil that can be obtained from the fractional distillation of crude oil and demand is the amount that is needed by consumers. The demand for petrol is far greater than the amount that can be supplied because it is a shorter chain hydrocarbon that can release energy more quickly by burning so making it a good fuel. In comparison the amount of diesel oil that can be supplied is far greater than the consumers' need for the fuel because it is a longer chain hydrocarbon and does not make as good a fuel as petrol. The shortfall in petrol supply can be overcome by taking the diesel oil that is in excess and cracking the diesel oil into shorter chain hydrocarbons such as petrol.

Unit C2

1. (a) Giant crystalline (or metallic) structure
(b) Sea of free electrons
(c) **Any three points, e.g.** Good conductor of heat; Good conductor of electricity; Have high melting and boiling points; Shiny, hard and strong; Ductile; Malleable

2. (a) Theoretical yield is calculated from relative formula masses of reactants and products **(1 mark)**; Actual yield is the actual mass of useful product obtained from a given amount of reactant in the experiment **(1 mark)**
(b) **Any suitable answer, e.g.** More economical; Cheaper process; Greater atom economy
(c) (i) $(\frac{31.5}{32.4}) \times 100 = 97.2\%$
(ii) RFM of ZnO = 65 + 16 = 81 **(1 mark)**; Percentage Zn in ZnO = $(\frac{65}{81}) \times 100 = 80.2\%$ **(1 mark)**
(iii) Mass of Zn = $(\frac{80.2}{100}) \times 31.5 = 25.3$g

3. (a) **Any suitable answer, e.g.** Because it gives out heat energy; The temperature of the reactants or surroundings increases/rises.
(b) Energy is needed to break the bonds in the reactants **(1 mark)**; Energy is given out when bonds are made in the products **(1 mark)**; In an exothermic reaction more energy is released in making bonds in the products than is needed to break bonds in the reactants **(1 mark)**
(c) Iodine **(1 mark)**; because chlorine is higher up the group / is more reactive **(1 mark)**

4. Helium is inert / doesn't react / is less dense than air so will float **(1 mark)**

5.*Iodine has a single covalent bond with one shared electron pair, and exists as a diatomic molecule. Because there are two atoms of iodine in each molecule it has a simple molecular structure. At room temperature iodine is a solid but it has a very low melting point because the intermolecular forces holding the diatomic molecules within their structure are weak. Weak intermolecular forces do not require much energy to overcome. This is why the melting points and boiling points of molecules with this type of structure are low – many are gases at room temperature. Iodine is unable to conduct electricity because a covalent bond does not involve the transfer of electrons or charged ions.

6. (a) Proton number is the same as the atomic number
(b) **(1 mark each)**; Mg = 12, N = 7, S = 16
(c) Mass number – atomic number = number of neutrons
(d) **(1 mark each)**

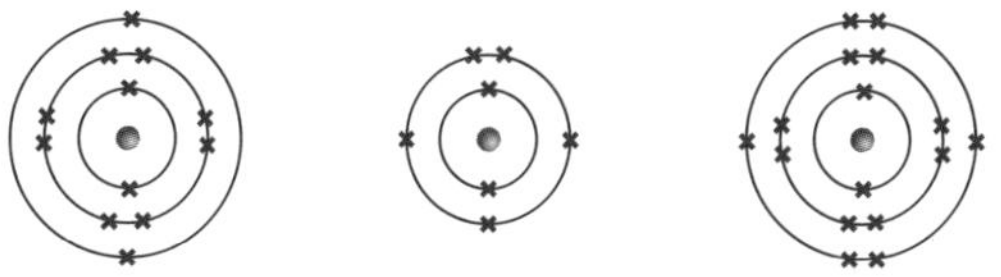

7. (a) **Any three from:** Catalyst; Temperature; Concentration; Surface area; Pressure (in gases)
(b) Increases the rate of reaction
(c) There are fewer particles and they are spread out **(1 mark)** so there will be less chance of collisions **(1 mark)**
(d) Compare the same mass of calcium carbonate chips to finely crushed calcium carbonate in the reaction with hydrochloric acid **(1 mark)**; The volume of carbon dioxide produced is measured every minute **(1 mark)**; The same volume of carbon dioxide produced will be reached faster with the finely crushed calcium carbonate **(1 mark)**; Finely crushed calcium carbonate has a larger surface area so increases the rate of reaction **(1 mark)**
(e) Reduces the minimum amount of energy **(1 mark)** needed for the reaction to happen **(1 mark)**; This means that there will be more successful collisions between the particles **(1 mark)**

8. (a) C
(b) C
(c) (i) $BaCl_2$ (ii) MgO (iii) $AlCl_3$ (iv) Fe_2O_3

9. (a) $Mg(s) + Cl_2(g) \longrightarrow MgCl_2(s)$ **(2 marks)**
(b) $4Fe(s) + 3O_2(g) \longrightarrow 2Fe_2O_3(s)$ **(2 marks)**

10. (a) (i) Fe_2O_3: (56 × 2) + (16 × 3) = 160
(ii) Al_2O_3: (2 × 27) + (3 × 16) = 102
(b) (i) 800g $Fe_2O_3 = \frac{800}{160} = 5$ formula masses **(1 mark)**; From equation, each Fe_2O_3 reacts with 2 Al, so need 2 × 5 = 10 Al **(1 mark)**; Mass of Al = 10 × 27 = 270g **(1 mark)**
(ii) 480g $Fe_2O_3 = \frac{480}{160} = 3$ formula masses **(1 mark)**; From equation, each Fe_2O_3 produces 2 Fe, so 3 formula masses of Fe_2O_3 produces 2 × 3 = 6 Fe **(1 mark)**; Mass of Fe = 6 × 56 = 336g **(1 mark)**
(iii) 612g $Al_2O_3 = \frac{612}{102} = 6$ formula masses **(1 mark)**; From equation, 2 Al is needed for each Al_2O_3, so need 2 × 6 = 12 Al **(1 mark)**; Mass of Al = 12 × 27 = 324g **(1 mark)**

11. (a) Diamond has a giant molecular structure **(1 mark)**; Each of the carbon atoms forms four covalent bonds with other carbon atoms, making it a very hard substance **(1 mark)**; It has a very high melting point and boiling point but it is unable to conduct electricity because there are no available electrons **(1 mark)**
(b) In graphite the carbon atoms are only covalently bonded to three further carbon atoms, leaving the fourth electron from each atom to move freely about the structure **(1 mark)**; This allows graphite to conduct electricity **(1 mark)**; The atoms within the structure of graphite are arranged in layers that are able to slide past each other. This allows graphite to be used in pencils and as a lubricant **(1 mark)**

Answers

Unit C3

1. **(a)** A homologous series has the same general formula / neighbouring molecules in a homologous series differ by CH_2 **(1 mark)**; Molecules of a homologous series have similar chemical properties / have a trend in physical properties **(1 mark)**

   (b) 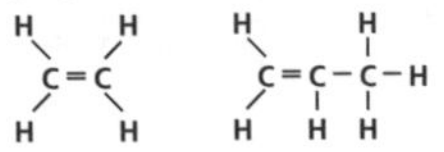

   (c) **Any three from:** Making solvents; Emollients; Esters; Emulsifiers; Thickeners; Beverages

2. **(a)** Oxidation is the loss of electrons from charged ions
   (b) Reduction is the gain of electrons from charged ions
   (c) Positively charged ions in the substance are attracted to the cathode where they gain electrons **(1 mark)**; This is reduction **(1 mark)**; At the same time negatively charged ions are attracted to the anode where they lose electrons **(1 mark)**; This is oxidation **(1 mark)**

3. **(a)** $C_6H_{12}O_6 \longrightarrow 2C_2H_5OH + 2CO_2$
   (b) Ferment contains a mixture of liquids with different boiling points **(1 mark)** so to obtain pure ethanol it must be distilled using fractional distillation **(1 mark)**
   (c) **(i)** An ester; **(ii)** Ethyl ethanoate

4. **(a)** Electrolysis
   (b) Active electrodes
   (c) They take part in the electrolysis process **(1 mark)** and either lose or gain mass **(1 mark)**
   (d) Metal ions / positive ions / cations **(1 mark)**; Because the cathode is negatively charged and attracts the metal ions **(1 mark)**

5. **(a)** RFM of HNO_3 = 1 + 14 + 3 × 16 = 63 **(1 mark)**; RFM of NH_4NO_3 = 14 + (4 × 1) + 14 + (3 × 16) = 80 **(1 mark)**
   (b) RFM of NH_4OH = 14 + (4 × 10) + 16 + 1 = 35 **(1 mark)**; $51 \times \frac{80}{35}$ =117 tonnes **(1 mark)**
   (c) $\frac{56}{80} \times 63$ = 44.1 tonnes **(2 marks)**
   (d) $\frac{(2 \times 14)}{80} \times 100$= 35% **(2 marks)**

6. **(a)** Minerals in hard water react with the soap **(1 mark)**; Some of the soap becomes insoluble as it is used up in the reaction **(1 mark)**; It is hard to form lather because scum is produced **(1 mark)**
   (b) A

7. **(a)** **1 mark for explanation, 1 mark for example:** An ester is made when a carboxylic acid is reacted with an alcohol. This is a reversible reaction. E.g. Ethanoic acid + Ethanol $\rightleftharpoons$ Ethyl ethanoate + Water
   (b) **Any three from:** Soluble in organic solvents; Have strong fruity smells; Solubility in water decreases as number of carbons in carbon chain increases; Vary from colourless volatile liquids to colourless waxy solids

8. **(a)** Room temperature and atmospheric pressure
   (b) **(i)** 1 mole of gas = $24dm^3$; 0.25mol = 24 × 0.25 = $6dm^3$
   (ii) 3 moles = 3 × 24 = $72dm^3$
   (iii) 0.5 moles = 0.5 × 24 = $12dm^3$
   (iv) 1.25 moles = 1.25 × 24 = $30dm^3$

9. *Industry will need to use more than one technique to alter the equilibrium of a reversible reaction because they are looking for the method that will produce the greatest possible yield of desired product in the shortest time possible that is also economically viable. If, for example, the reaction used by the industry was an exothermic reaction in the forwards direction, then increasing the temperature of the reaction would shift the equilibrium to favour the reactants. So although a higher temperature increases the rate of reaction, because the reaction is a reversible reaction, the equilibrium will shift to compensate. This results in decomposition of the product and less yield produced, as the dynamic equilibrium is restored. Catalysts are used to increase the speed at which the reaction responds to changes in the conditions. A change in concentration (or pressure, if gases) of reactants enables the industry to have some control over the position of the equilibrium of the reaction and therefore to gain a product at the best yield and most economical costs.

10. **(a)** It goes to the side with the smaller number of molecules.
    (b) Give out more heat.
    (c) The reaction reaches dynamic equilibrium.
    (d) Equilibrium will shift **(1 mark)** to cancel out the change **(1 mark)**

11. **(a)** Number of moles = concentration × volume **(1 mark)**; Moles = 0.4 × 0.75 = 0.3mol **(1 mark)**
    (b) Mass of solute = RFM × moles **(1 mark)**; Mass of $CuCl_2$ = (63.5 + 71) × 0.3 = 40.35g **(1 mark)**
    (c) Concentration = moles ÷ volume **(1 mark)**; Concentration = 0.3 ÷ 0.25 = 1.2mol dm^{-3} **(1 mark)**
    (d) Mass solute = RFM × moles **(1 mark)**; Mass $CuSO_4$ = (63.5 + 32 + 64) × 0.3 = 47.85g **(1 mark)**

12. Sulfur dioxide **(1 mark)**; Avogadro's law states that equal volumes of gases at the same temperature and pressure contain the same number of particles **(1 mark)**; Sulfur dioxide occupies the greater volume so it must contain more particles **(1 mark)**

13. 1 mole of H_2SO_4 neutralises 2 moles of NH_4OH
    Moles of H_2SO_4: 0.8mol dm^{-3} × $0.025dm^3$ = 0.02mol **(1 mark)**
    Moles of NH_4OH used up: 2 × 0.02 = 0.04mol **(1 mark)**
    Concentration of NH_4OH: $\frac{\text{moles}}{\text{volume}}$ = 0.04 ÷ 0.15 = 0.267mol dm^{-3} **(1 mark)**

Glossary

Acid rain – rain that has reacted with gaseous pollutants such as sulfur dioxide and nitrogen dioxide. The gases are as a result of fossil fuels being burned.

Active electrode – an electrode that takes part in the electrolysis reaction.

Actual yield – the amount of wanted product obtained from a chemical reaction.

Addition polymerisation – reaction in which a single type of monomer gives rise to a single polymer, with no other reaction products.

Alkali metal – an element found in Group 1 of the periodic table. Atoms of these elements all contain a single electron in the outer energy level.

Alkane – a saturated hydrocarbon that contains only single carbon–carbon covalent bonds. The number of hydrogen atoms is double the number of carbon atoms plus two.

Alkene – an unsaturated hydrocarbon that contains one double covalent carbon–carbon bond. The number of hydrogen atoms is double the number of carbon atoms.

Alloy – a mixture of two or more metals, or a mixture of one metal and a non-metal.

Ammonia gas – product of the Haber process made from nitrogen and hydrogen and used to manufacture nitric acid and fertilisers.

Anion – a negatively charged particle formed when an atom or group of atoms lose or gain electrons.

Anode – the electrode connected to the positive terminal of a battery.

Argon – noble gas with the symbol Ar.

Atom – smallest particle of a chemical element that can exist.

Bacteria – group of microscopic, single-celled organisms that inhabit virtually all environments, including soil and water.

Balanced symbol equations – the reaction written as formulae where there is an equal number of total atoms of reactant to product.

Biofuel – a source of renewable energy that is made from biological materials that include plants and animal waste.

Butane – alkane with the formula C_4H_8 that is used as a fuel.

Calcium carbonate – compound that has the formula $CaCO_3$ and is found in many different types of rock.

Calcium hydrogen carbonate – compound that has the formula $Ca(HCO_3)_2$ and is found dissolved in water. Also known as calcium bicarbonate.

Calcium ions – formed when calcium metal atoms lose electrons during chemical reactions.

Catalyst – a substance that is used to speed up a chemical reaction without itself being used. At the end of the reaction, a catalyst will not be chemically changed.

Cathode – the electrode connected to the negative terminal of the battery.

Cation – a positively charged particle formed when an atom or group of atoms lose electrons.

Cement – a building material made by grinding calcined limestone and clay to a fine powder, then heating in a rotary kiln.

Chemical reaction – a process that leads to the transformation of one set of chemical substances to another.

Chloride ions – chlorine atoms that have gained an additional electron as a result of chemical reaction.

Chlorine – halogen with symbol Cl; a yellow, poisonous gas.

Chromatography – a technique used to separate unknown mixtures for analysis.

Compound – a substance that contains two or more different atoms of one or more elements chemically combined.

Compound ion – a positively or negatively charged particle formed when a group of atoms lose or gain electrons.

Concentration – the amount of atoms or particles within a fixed or known volume.

Concrete – used in the construction industry and made by combining cement with sand, aggregate and water.

Cosmetic – general term used to describe a preparation used externally to condition and beautify the body.

Covalent bond – the sharing of electron pairs between non-metals.

Cracking – the process of breaking down larger hydrocarbons into smaller, more useful ones using a catalyst and heat.

Crude oil – a naturally occurring flammable liquid that is made up of a mixture of hydrocarbons.

Decolourise – to lose colour, for example, bromine water will lose its colour when added to an alkene.

Deforestation – the clearance of naturally occurring forests by logging and burning.

Detergent – a substance that in solution aids the removal of dirt and other foreign matter from contaminated surfaces.

Discharged – in electrolysis this is when the positive or negative ion either gains or loses electron at the electrode to become a neutral atom.

Displacement reaction – when elements of greater reactivity within a compound swap places with elements of lesser reactivity.

Dissolved substances – minerals and other chemicals that are dissolved in water or other solvents.

Glossary

Double bond – the sharing of two pairs of electrons between non-metals.

Double covalent carbon–carbon bonds – when two carbon atoms share two pairs of electrons.

Electrolysis – splitting up of a molten or aqueous solution of salt, using electricity.

Electrolyte – an aqueous or molten substance that contains free moving ions and as a result is able to conduct electricity.

Electronic configuration – the orbital/shell arrangement of electrons around the nucleus of an atom.

Electron – a negatively charged particle that orbits the nucleus of an atom. It has a negligible mass.

Element – a pure substance that is made from just one type of atom. It cannot be chemically split into anything simpler.

Emollient – substance used to soothe and soften the skin.

Empirical formula – the simplest formula of a compound.

Emulsifier – an additive that will stop two liquids from separating when they are mixed together, especially if those liquids would not normally mix together.

Endothermic reaction – a chemical reaction that will give out energy to the surroundings in the form of heat.

Energy – the ability to do work.

Ester – an organic compound formed when an organic acid reacts with an alcohol.

Ethane – a saturated hydrocarbon with formula C_2H_6, ethane is a colourless, odourless flammable gaseous alkane obtained from natural gas and crude oil. It is used as a fuel and in the manufacture of organic chemicals.

Ethanol – a clear, colourless alcohol found in beverages such as wine, beer and brandy. It has the formula C_2H_5OH.

Ethene – an unsaturated hydrocarbon with the formula C_2H_4, ethene is a colourless, flammable, gaseous alkene obtained by the cracking of crude oil and the dehydration of alcohol. It is the simplest alkene and is used in the manufacture of organic chemicals.

Exothermic reaction – a chemical reaction that will take in energy from the surroundings in the form of heat.

Expected yield – the amount of product that should be produced from a given amount of reactants where one of the reactants is a limiting reagent.

Extrusive igneous rock – fast cooling of lava at the Earth's surface forms rock with small crystals.

Fatty acid – the common name for a group of organic acids.

Flavouring – an edible chemical and/or extract that alters the flavour of food and food products.

Formula – shows the number of atoms of each element present in a chemical compound.

Fossil fuel – natural source of energy, such as oil, coal or natural gas, that has been made from the remains of plants and animals over millions of years.

Fractional distillation – a method of separating a mixture of liquids that have different boiling points.

Fractions – a group of hydrocarbons with similar boiling points.

Glass – a non-crystalline solid material made from heating limestone and silicon dioxide.

Global warming – an increase in the Earth's average atmospheric temperature that causes corresponding changes in climate and that may result from the greenhouse effect.

Group – a vertical column of the periodic table in which elements have similar properties.

Halide – compound formed from the chemical reaction of a halogen.

Halide ion – formed when elements of Group 7 gain electrons during a chemical reaction.

Halogens – non-metals found in Group 7 of the periodic table. Known as the 'salt-formers', they exist as diatomic molecules.

Hard water – water that has a high mineral content from the dissolved calcium and magnesium carbonates.

Hydrocarbons – compounds that are made of only hydrogen atoms and carbon atoms.

Hydroxide – a compound containing a hydroxyl ion (OH^-).

Identified – the process of having found out the components of a substance.

Igneous rock – rock that is formed by the cooling and solidification of molten magma.

Immiscible – substances that will not mix together but form two or more distinctly separate layers.

Impurities – foreign elements that can have an effect on properties of the substance.

Insoluble – a substance that will not dissolve in a solvent and is unable to form a solution.

Intrusive igneous rock – rocks with large crystals formed by the slow cooling of molten magma within the Earth's crust.

Ion – a positively or negatively charged particle formed when an atom or group of atoms lose or gain electrons.

Ionic bond – a chemical bond in which one atom loses an electron to form a positive ion and the other atom gains an electron to form a negative ion.

Krypton – noble gas with the symbol Kr.

Law of conservation of mass – in chemical reactions mass is neither lost or gained.

Mendeleev, Dmitri – Russian chemist who devised the arrangement of the modern periodic table.

Metal atom – an atom of an element on the left-hand side of the periodic table.

Metal compound – a substance that contains atoms of one or more elements chemically combined, one of which will be a metal.

Metal ions – formed when metal atoms lose electrons.

Metal oxides – formed when a metal ion combines with an oxide ion.

Metamorphic rock – sedimentary or igneous rock that has been changed by the action of intense heat and pressure.

Mixture – two or more different substances that are not chemically combined.

Molar mass – is the mass of one mole of a substance.

Molecule – the smallest part of an element or compound that can exist on its own.

Monomer – a carbon-based molecule that can combine with others to form a polymer.

Monounsaturated – when a long carbon chain molecule contains only one double bond.

Negatively charged ions – a charged particle formed when an atom or group of atoms gain electrons.

Neon – noble gas with the symbol Ne.

Neutralisation – when an alkali is reacted with an acid to form a salt and water. The resultant solution will have a pH of 7.

Neutron – uncharged particle found in the nucleus of almost all atoms.

Nitrates – compounds found in fertilisers used by farmers; can have a negative impact on water.

Noble gases – unreactive non-metallic elements found in Group 0 of the periodic table.

Non-metal atom – an atom of an element found on the right-hand side of the periodic table.

Non-metal ions – formed when non-metal atoms gain electrons.

Odour – a distinctive smell given off by a chemical.

Ore – a type of rock that contains minerals with important elements including metals in sufficient quantities to make it commercially viable to extract.

Organic acid – a carbon-based compound acid.

Oxidation – when electrons are lost from a reacting element or compound.

Periodic table – a tabular display of all known discovered elements.

Period – a horizontal row of elements with a variety of properties and the same number of electron shells.

Pesticides – chemical compounds formulated to kill insect pests, for example aphids on farmers' crops.

Plaster – a mixture of lime, sand and water, sometimes with fibre added, that hardens to a smooth solid and is used for coating walls and ceilings.

Poly(chloroethene) – made from chlorinated ethene molecule and commonly referred to as PVC. It can be used to make windows and doors.

Poly(ethene) – made from the monomer ethene and used to make plastic bags.

Polymer – a large molecule composed of repeating structural units, typically connected by covalent chemical bonds.

Poly(propene) – made from the monomer propene and used to make plastic bottles.

Polyunsaturated – when long carbon chain molecule contains more than one double bond.

Positively charged ion – a charged particle formed when an atom or group of atoms lose electrons.

Precipitate – the name given to the solid that is formed as a result of a precipitation reaction.

Precipitation reaction – the formation of an insoluble solid when two solutions are mixed.

Products – the substances present at the end of a chemical reaction.

Propane – a saturated hydrocarbon with the formula C_3H_8, propane is a colourless gas, found in natural gas and crude oil. It is widely used as a fuel.

Propene – has the formula C_3H_6 and contains one double carbon–carbon bond.

Proton – a positively charged particle found in the nucleus of every atom.

PTFE – polymer made from fluorinated ethene that has non-stick properties. It is commonly called Teflon.

Pure – a substance that is free from foreign elements.

Qualitative analysis – a method used to establish the presence of a given element, compound or functional group.

Quantitative analysis – a method used to establish the amount of a given element or compound in a sample.

Reactants – the substances present at the beginning of a chemical reaction.

Reactivity series – an order of reactivity of metals with metal of highest reactivity placed at the top.

R_f value – in chromatography, the distance travelled by a given component divided by the distance travelled by the solvent front.

Reduction – the process in which oxygen is lost, or in which hydrogen or electrons are gained.

Glossary

Relative atomic mass – the average mass of an atom compared with the mass of an atom of carbon.

Relative formula mass – the sum of the relative atomic masses as shown in a chemical formula.

Reversible – a chemical reaction that is able to react in both directions.

Saponification – the reaction between oils and sodium or potassium hydroxide that results in the formation of soap.

Saturated – hydrocarbon molecules that are completely surrounded with hydrogen and have no further available bond sites.

Sedimentary rock – rock that is formed by consolidated sediment deposited in layers.

Single covalent bond – the sharing of a single pair of electrons between non-metals.

Single covalent carbon–carbon bonds – when two carbon atoms share one pair of electrons.

Soap – a cleansing agent made from animal and vegetable fats, oils and greases.

Sodium chloride – an ionic compound from the reaction of sodium metal and chlorine non-metal.

Soft water – water that has a higher level of sodium ions than other minerals.

Solid substances – in water analysis, substances that are suspended in or float on water.

Solubility – the ability of a solid to dissolve in a solvent to make a solution.

Soluble – a substance that will dissolve in a solvent to form a solution.

Solute – a substance (solid) that will dissolve in another to form a solution.

Solvent – a substance (liquid) that dissolves another to form a solution.

Sulfate ions – a compound ion made of sulfur and oxygen.

Sustainable development – a pattern of use of resources that aims to meet the needs of the population, while preserving the environment so that these needs can be met not only for the present, but also for future generations.

Temperature – a measure of the amount of heat energy.

Temporary hardness – hardness in water caused by a high mineral content from dissolved calcium hydrogen carbonate and/or magnesium hydrogen carbonate that can be removed by boiling.

Theoretical yield – the maximum amount of wanted product that can be obtained from a chemical reaction.

Thermal decomposition – the breaking down of a large molecule to form smaller molecules by using heat.

Thickeners – substances that, when added to an aqueous mixture, increase its viscosity without substantially modifying its other properties, such as taste.

Titration – a method to measure how much of one solution is needed to react exactly with another of known volume.

Transition metals – found in the middle of the periodic table.

Unsaturated – when a long carbon chain molecule contains double bonds.

Unsaturated hydrocarbons – hydrocarbon molecules that contain one or more carbon–carbon double bond.

Waxy solid – a solid substance that has the appearance and texture of wax.

Xenon – noble gas with the symbol Xe.

Zinc ions – formed when zinc metal atoms lose electrons during chemical reactions.

HT

Alcohol – a carbon-based compound that contains a hydroxyl (–OH) group.

Avogadro's law – under identical conditions of temperature and pressure, equal volumes of gases contain an equal number of molecules.

Carboxylic acid – a carbon-based compound that contains the group –COOH.

Dynamic equilibrium – when the rate of a reversible reaction occurs at the same rate in both directions.

Hydrogenated – when a large fatty acid has had additional hydrogen added to its molecule.

Hydrophilic – used to describe a substance that will attract, dissolve in or absorb water.

Hydrophobic – used to describe a substance that repels or hates water.

Isotopes – different types of atom of the same chemical element, each having a different number of neutrons.

Molar volume – is the volume occupied by one mole of a substance (chemical element or chemical compound) at a given temperature and pressure.

Mole – a measure of the amount of a substance in terms of the number of particles it contains.

Nitinol – a shape memory alloy of nickel and titanium.

Surface tension – a force acting within the surface of a liquid that makes the liquid behave as if it were covered in a stretched elastic skin.

Key

relative atomic mass **atomic symbol** name atomic (proton) number

1	2											3	4	5	6	7	0
							1 **H** hydrogen 1										4 **He** helium 2
7 **Li** lithium 3	9 **Be** beryllium 4											11 **B** boron 5	12 **C** carbon 6	14 **N** nitrogen 7	16 **O** oxygen 8	19 **F** fluorine 9	20 **Ne** neon 10
23 **Na** sodium 11	24 **Mg** magnesium 12											27 **Al** aluminium 13	28 **Si** silicon 14	31 **P** phosphorus 15	32 **S** sulfur 16	35.5 **Cl** chlorine 17	40 **Ar** argon 18
39 **K** potassium 19	40 **Ca** calcium 20	45 **Sc** scandium 21	48 **Ti** titanium 22	51 **V** vanadium 23	52 **Cr** chromium 24	55 **Mn** manganese 25	56 **Fe** iron 26	59 **Co** cobalt 27	59 **Ni** nickel 28	63.5 **Cu** copper 29	65 **Zn** zinc 30	70 **Ga** gallium 31	73 **Ge** germanium 32	75 **As** arsenic 33	79 **Se** selenium 34	80 **Br** bromine 35	84 **Kr** krypton 36
85 **Rb** rubidium 37	88 **Sr** strontium 38	89 **Y** yttrium 39	91 **Zr** zirconium 40	93 **Nb** niobium 41	96 **Mo** molybdenum 42	[98] **Tc** technetium 43	101 **Ru** ruthenium 44	103 **Rh** rhodium 45	106 **Pd** palladium 46	108 **Ag** silver 47	112 **Cd** cadmium 48	115 **In** indium 49	119 **Sn** tin 50	122 **Sb** antimony 51	128 **Te** tellurium 52	127 **I** iodine 53	131 **Xe** xenon 54
133 **Cs** caesium 55	137 **Ba** barium 56	139 **La*** lanthanum 57	178 **Hf** hafnium 72	181 **Ta** tantalum 73	184 **W** tungsten 74	186 **Re** rhenium 75	190 **Os** osmium 76	192 **Ir** iridium 77	195 **Pt** platinum 78	197 **Au** gold 79	201 **Hg** mercury 80	204 **Tl** thallium 81	207 **Pb** lead 82	209 **Bi** bismuth 83	[209] **Po** polonium 84	[210] **At** astatine 85	[222] **Rn** radon 86
[223] **Fr** francium 87	[226] **Ra** radium 88	[227] **Ac*** actinium 89	[261] **Rf** rutherfordium 104	[262] **Db** dubnium 105	[266] **Sg** seaborgium 106	[264] **Bh** bohrium 107	[277] **Hs** hassium 108	[268] **Mt** meitnerium 109	[271] **Ds** darmstadtium 110	[272] **Rg** roentgenium 111	Elements with atomic numbers 112–116 have been reported but not fully authenticated						

*The lanthanoids (atomic numbers 58–71) and the actinoids (atomic numbers 90–103) have been omitted.
The relative atomic masses of copper and chlorine have not been rounded to the nearest whole number.

Notes

Notes

Index

Index